SpringerBriefs in Electrical and Computer Engineering

SpringerBriefs present concise summaries of cutting-edge research and practical applications across a wide spectrum of fields. Featuring compact volumes of 50 to 125 pages, the series covers a range of content from professional to academic. Typical topics might include: timely report of state-of-the art analytical techniques, a bridge between new research results, as published in journal articles, and a contextual literature review, a snapshot of a hot or emerging topic, an in-depth case study or clinical example and a presentation of core concepts that students must understand in order to make independent contributions.

More information about this series at http://www.springer.com/series/10059

Hugo Alexandre de Andrade Serra • Nuno Paulino

Design of Switched-Capacitor Filter Circuits using Low Gain Amplifiers

 Springer

Hugo Alexandre de Andrade Serra
Faculdade de Ciências e Tecnologia
UNINOVA
Caparica
Portugal

Nuno Paulino
Department of Electrical Engineering
UNINOVA
Caparica
Portugal

ISSN 2191-8112   ISSN 2191-8120 (electronic)
SpringerBriefs in Electrical and Computer Engineering
ISBN 978-3-319-11790-4   ISBN 978-3-319-11791-1 (eBook)
DOI 10.1007/978-3-319-11791-1

Library of Congress Control Number: 2014954623

Springer Cham Heidelberg New York Dordrecht London

Printed on acid-free paper

Springer is part of Springer Science+Business Media (www.springer.com)

Preface

Analog filters arc extremely important blocks in several electronic systems such as RF transceivers or sigma delta modulators. They allow selecting between signals with different frequency and eliminating unwanted signals.

In modern deep-submicron CMOS technologies the intrinsic gain of the transistors is low and has a large variability, making the design of moderate and high gain amplifiers extremely difficult.

The objective of this book is to study switched-capacitor (SC) circuits based on the low-pass and band-pass Sallen-Key topologies, since they do not require high gain amplifiers. The strategy used to achieve this objective is to replace the operational amplifier (opamp) with a voltage buffer. Doing this simplifies the design of the amplifier although it also eliminates the virtual ground node from the circuit. Without this node parasitic insensitive SC networks cannot be used. Due to modern parasitic extraction software that can reliably predict the values of parasitic capacitances, the historical disadvantage of parasitic sensitive SC networks (parallel SC) is no longer critical, allowing its influence to be compensated during the design process.

Different types of switches were simulated to determine the one with the least non-linear effects. Two techniques (common mode voltage adjustment and source degeneration) were used to reduce the distortion introduced by the buffers.

Low-pass (second and sixth order) and band-pass (second and fourth order) SC filters were simulated in differential configuration in standard 130 nm CMOS technology, having obtained for the low-pass filter a distortion of -62 dB for the biquad section and -54 dB for the sixth-order filter, for a cutoff frequency of 1 MHz and when operating at 100 MHz of clock frequency. The total power consumption was 986 μW (i.e. 986×10^{-6}) and 5.838 mW, respectively.

Acknowledgements

This work was supported by projects DISRUPTIVE (EXCL/EEI-ELC/0261/2012) and PEST (PEst-OE/EEI/UI0066/2014).

Contents

List of Abbreviations and Acronyms

A/D	Analog/Digital
CDFA	Complementary Differential Folded Amplifier
CMOS	Complementary Metal-Oxide-Semiconductor
CSF	Complementary Source Follower
D/A	Digital/Analog
DFA	Differential Folded Amplifier
FOM	Figure of Merit
IC	Integrated Circuit
KCL	Kirchhoff's Current Law
KVL	Kirchhoff's Voltage Law
MOSFET	Metal-Oxide-Semiconductor Field-Effect Transistor
OpAmp	Operational Amplifier
SC	Switched-Capacitor
SF	Source Follower
SFG	Signal Flow Graph
VCVS	Voltage Controlled Voltage Source

Chapter 1
Introduction

Abstract This chapter gives a brief description of SC circuits and the motivation behind the work. The book organization per chapter is also described.

Interest in switched-capacitor networks started in the late 70 s due to the possibility of implementing analog filters using monolithic integrated circuit (IC) technology and because it is possible to obtain a good accuracy in the ratio between two capacitor values. Since typically high value resistors are needed in the implementation of analog filters, they can be generated using small on-chip capacitors, which makes it an attractive option for monolithic fabrication, since it occupies a small substrate area.

SC circuits operate as discrete-time circuits without the use of converters (A/D or D/A). When used as filters, these circuits have an accurate frequency response, good linearity, and good dynamic range. The accuracy of these circuits comes from the time constants being determined by capacitor ratios, which in real applications can have an accuracy of nearly 0.1 % while in integrated RC circuits the time constant error can range from 20 to 50 %. Another advantage is the incorporation of a certain degree of frequency tuning by varying the clock frequency [1].

Although there are several advantages in using this type of technology, there are also several non-ideal properties that need to be considered, like clock feedthrough, offset error, noise, and parasitic capacitances [2]. Usually SC filters need an antialiasing filter at the input and a smoothing filter at the output [3].

The scaling-down of transistors in advanced deep-submicron CMOS technologies results in the reduction of the intrinsic gain (g_m/g_{ds}) [4], making the design of high gain opamps with high bandwidth increasingly difficult. This limitation has large impact on the performance of filter circuits.

The objective of this book is to study the low-pass and band-pass Sallen-Key topologies since they do not require high gain amplifiers. The strategy used is to replace the opamp with a voltage buffer. Doing this simplifies the design of the amplifier although it also eliminates the virtual ground node from the circuit. Without this node, parasitic insensitive SC networks cannot be used. Due to modern parasitic extraction software that can reliably predict the values of parasitic capacitances, the historical disadvantage of parasitic sensitive SC networks (parallel SC) is no longer critical, allowing its influence to be compensated during the design process.

© Springer International Publishing Switzerland 2015
H. A. de A. Serra, N. Paulino, *Design of Switched-Capacitor Filter Circuits using Low Gain Amplifiers,* SpringerBriefs in Electrical and Computer Engineering, DOI 10.1007/978-3-319-11791-1_1

Since the first SC circuits used parallel SC networks, the filter topologies studied in this book will be converted to SC circuits using this type of network, and the topologies adapted, if need be, to work in modern nm CMOS technologies.

Besides this introductory chapter, this book is organized in six more chapters.

In the second chapter, an overview of SC filters and their main building blocks is presented, including advantages and non-ideal properties that need to be considered in this type of circuits. SC resistor emulation networks are also presented and an integrator is described using two different networks in order to show the advantage of using certain SC networks to neglect non-ideal properties when an amplifier is used. The transfer functions of the low-pass and band-pass Sallen-Key topologies are also presented in this chapter.

In the third chapter, the low-pass near-unity gain Sallen-Key topology is presented along with the design equations and transfer function. Using this topology and adapting it, a SC single-ended version is obtained. The design equations and transfer function are also presented, and the topology is simulated using ideal components. The differential configuration of the low-pass SC filter is presented. A sixth-order filter is also obtained by cascading biquadratic sections and simulated using ideal components.

The fourth chapter has the same structure as the previous chapter, but a band-pass filter is considered and a fourth-order filter is obtained from cascading biquadratic sections.

In the fifth chapter, the non-ideal effects that result from using real components are described. The non-linear effects due to real switches is studied using different types of switches in a first-order SC filter and varying the switches size. Two clock boost circuits, which can be used to reduce the non-linear effects of switches, are presented, and the switches conductance with and without the use of these circuits is compared. One of these circuits is capable of operating at lower voltages. Two different buffers are also presented, in normal and in complementary configurations, along with low frequency gain, medium frequency gain and noise equations.

In the sixth chapter, the filter circuits are simulated using real components. The results of the simulations are presented when all components are ideal, when the switches are real and the remaining components ideal, when the buffers/amplifiers are real and the remaining components ideal, and when all components are real in order to determine the influence of each real component in the overall performance of the filter. The circuits include second-order low-pass filters, two operating at 1.2 V using the buffers presented in the previous chapter, one operating at 0.9 V, and a sixth-order filter obtained from cascading three biquadratic sections operating at 1.2 V. The band-pass filter circuit was also simulated for a second-order filter, and a fourth-order filter obtained from cascading two biquadratic sections, all operating at 1.2 V.

Finally, in the seventh chapter, the results obtained are discussed and the conclusions of this book are presented.

Chapter 2
Switched-Capacitor Circuits

Abstract This chapter introduces SC circuits. A brief description is given for the main building blocks of a SC filter (operational amplifiers, switches, capacitors, and non-overlapping clock phases). A few examples of SC resistor emulation circuits are presented along with their equivalent resistance. Two examples of SC integrators are also shown, one with the resistors implemented with a T branch (parasitic-sensitive integrator) and another with a branch in π (parasitic-insensitive integrator). A signal flow graph is presented which allows large circuits to be analyzed graphically. The chapter ends with the two Sallen-Key topologies designed in this book, in their continuous time version.

2.1 Switched-Capacitor Filters Building Blocks

SC filters can be implemented using switches, capacitors, opamps, and non-overlapping clock generators.

2.1.1 Operational Amplifiers

Opamps in SC circuits provide a virtual ground node. Parasitic capacitances connected to this node do not influence the performance of the circuit, since they will be connected to ground during one clock phase and to the virtual ground during the other. Opamps also have non-ideal effects that affect the performance of SC circuits like DC gain, unity gain frequency and phase margin, slew-rate, and common mode voltage.

2.1.2 Switches

SC circuits require switches with high off resistance in order to minimize the charge leakage when the switches are open, and low on resistance so that the circuit can settle in less than half a clock period. MOSFET transistors satisfy both these requirements, since they have very high off resistance and low on resistance ($G\Omega$ and $k\Omega$,

© Springer International Publishing Switzerland 2015
H. A. de A. Serra, N. Paulino, *Design of Switched-Capacitor Filter Circuits using Low Gain Amplifiers*, SpringerBriefs in Electrical and Computer Engineering, DOI 10.1007/978-3-319-11791-1_2

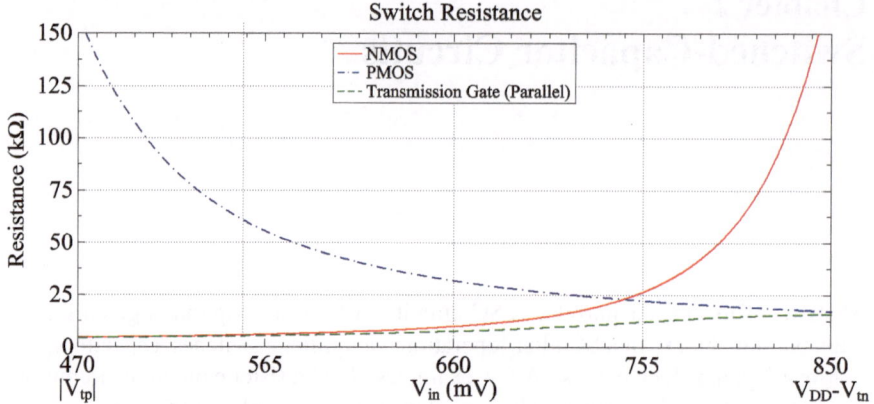

Fig. 2.1 Example of MOS transistors r_{ds} resistance as function of the $V_{in}(V_s)$ voltage

Fig. 2.2 NMOS switch
model considering source and
drain parasitic capacitances

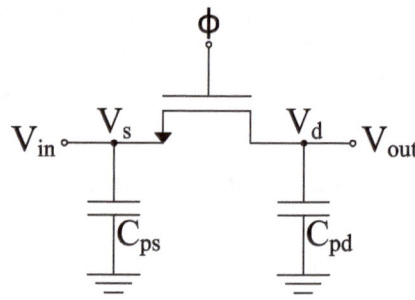

respectively), depending on the transistors size. Increasing the width of the transistor will decrease the value of both of these resistances. Switches can be implemented using NMOS transistors, PMOS transistors, or both in parallel (transmission gate). Using both in parallel will decrease not only the resistance value of the switch but also improve its linearity (Fig. 2.1). While NMOS transistors are better at conducting lower voltages (logic value '0') since they stop conducting when the input voltage is close to $V_{DD} - V_{tn}$, PMOS transistors are better at conducting higher voltages (logic value '1') since they do not conduct until the voltage is higher than V_{tp}. Transmission gates are good at transmitting '0's and '1's since lower voltages will travel via the NMOS transistor and higher voltages will travel via the PMOS transistor.

Depending on the SC circuit, it may be necessary to consider the switch model including parasitics (Fig. 2.2), since they may influence the transfer function of the SC filter. Examples of SC networks that are sensitive and insensitive to these parasitic capacitances are given in Sect. 2.2. While the resistance of a switch will decrease with the increase in size of the switch, the parasitic capacitances increases. It is important to note that the parasitic capacitances of a MOS transistor are non-linear, i.e., their value varies with the applied voltage value.

Fig. 2.3 Two-phase clock
generator

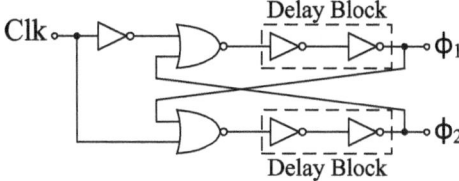

2.1.3   Capacitors

Capacitors are another element for which, depending on the SC network used, it may be necessary to consider the parasitic capacitances. The bottom plate parasitic capacitance can be as high as 20 % of the capacitance value, while the top plate parasitic capacitance can be as high as 5 % [1].

2.1.4   Non-Overlapping Clock Phases

The switches present in SC circuits require at least a pair of non-overlapping clock phases to preform the charge transfer. It is required that the phases do not overlap, so that no charge is accidentally lost by having two switches closed at the same time. An example of a two-phase clock generator is shown in Fig. 2.3.

2.2   Switched-Capacitor Resistor Emulation Networks

SC circuits emulate resistors using a combination of switches and capacitors. In order to obtain the equivalent resistance it is necessary to first calculate the average current value that flows from the input into the circuit. Considering the parallel network (Table 2.1), in which the input current only flows into the circuit during half a period ($0 \leq t \leq T/2$),

$$i_{rms} = \frac{1}{T} \int_0^{T/2} i_C(t)\, dt \tag{2.1}$$

Since the relation between charge and current is $i(t) = dq(t)/dt$,

$$i_{rms} = \frac{1}{T} \int_0^{T/2} dq_C(t) = \frac{Q_C(T/2) - Q_C(0)}{T} \tag{2.2}$$

Considering the phase scheme shown in Fig. 2.4, and that at $t = 0$ the capacitor maintains the voltage from the last phase, then at $t = 0 = T$ phase 2 is active and at $t = T/2$ phase 1 is active.

Fig. 2.4 Non-overlapping
clock phase scheme

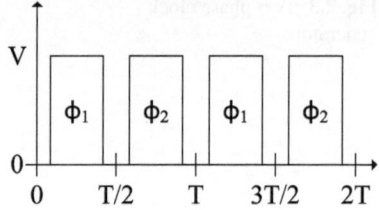

Using the charge values presented in Table 2.1 for the parallel network,

$$i_{rms} = \frac{(V_{in} - V_{out})C}{T} \tag{2.3}$$

Considering that the average current that flows through the resistance,

$$i_{rms} = \frac{V_{in} - V_{out}}{R} \tag{2.4}$$

By equating both equations (Eqs. 2.3 and 2.4) the equivalent resistance for the
parallel network is obtained.

$$R_{eq} = \frac{T}{C} \tag{2.5}$$

For the networks where the current flows into the network in both phases (series-
parallel and bilinear), the average current calculation must contemplate both phases.
For the series-parallel network (Table 2.1),

$$i_{rms} = \frac{1}{T}\left(\int_0^{T/2} dq_{C_2}(t) + \int_{T/2}^{T} dq_{C_1}(t)\right) \tag{2.6}$$

$$= \frac{Q_{C_2}(T/2) - Q_{C_2}(0)}{T} + \frac{Q_{C_1}(T) - Q_{C_1}(T/2)}{T}$$

Replacing the charge variables with the corresponding values that are shown in
Table 2.1,

$$i_{rms} = \frac{(V_{in} - V_{out})C_2}{T} + \frac{(V_{in} - V_{out})C_1 - 0}{T} \tag{2.7}$$

and by equating Eqs. 2.7 and 2.4, the equivalent resistance for the series-parallel
network is obtained.

$$R_{eq} = \frac{T}{C_1 + C_2} \tag{2.8}$$

Table 2.1 shows a few examples of circuits that can emulate resistors, their
equivalent resistance, and the charge in the capacitor(s) in each phase.

Table 2.1 SC resistor emulation circuits [5]

Circuit	Schematic	R_{eq}	$Q(\varphi_1)$	$Q(\varphi_2)$
Parallel	$V_{in} \overset{\phi_1}{} \overset{\phi_2}{} V_{out}$, C	$\dfrac{T}{C}$	$V_{in}C$	$V_{out}C$
Series	$V_{in} \overset{\phi_1}{} \overset{\phi_2}{} V_{out}$, C	$\dfrac{T}{C}$	0	$(V_{in} - V_{out})C$
Series-Parallel	$V_{in} \overset{\phi_1}{} \overset{\phi_2}{} V_{out}$, C_1, C_2	$\dfrac{T}{C_1 + C_2}$	0 $V_{in}C_2$	$(V_{in} - V_{out})C_1$ $V_{out}C_2$
Bilinear	$V_{in} V_{out}$, ϕ_1 ϕ_2 C ϕ_2 ϕ_1	$\dfrac{1}{4}\dfrac{T}{C}$	$(V_{in} - V_{out})C$	$(V_{out} - V_{in})C$

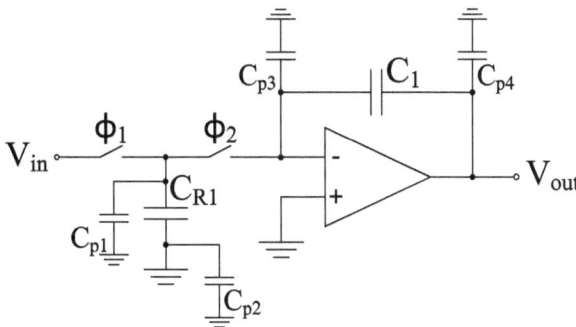

Fig. 2.5 Switched-capacitor parasitic-sensitive integrator

2.2.1 Parasitic-Sensitive Integrator

The SC parasitic-sensitive integrator (Fig. 2.5) was suggested in [6] by replacing the resistor in a Miller integrator with a parallel switch network.

Analyzing the charge across the capacitors in each phase, and considering that the output of the circuit will be sampled at the end of phase ϕ_1, Table 2.2 is obtained. Capacitance C_{p1} represents the top plate parasitic capacitance of C_{R1} and the parasitic

Table 2.2 Charge in the
capacitors in each phase

	$(n-1)T$	$(n-0.5)T$	nT
$Q_{C_{R1}}$	$V_{in}[(n-1)T]C_{R1}$	0	$V_{in}[nT]C_{R1}$
Q_{C_1}	$-V_{out}[(n-1)T]C_1$	$-V_{out}[(n-0.5)T]C_1$	$-V_{out}[nT]C_1$
$Q_{C_{p1}}$	$V_{in}[(n-1)T]C_{p1}$	0	$V_{in}[nT]C_{p1}$
$Q_{C_{p2}}$	0	0	0
$Q_{C_{p3}}$	0	0	0
$Q_{C_{p4}}$	$V_{out}[(n-1)T]C_{p4}$	$V_{out}[(n-0.5)T]C_{p4}$	$V_{out}[nT]C_{p4}$

capacitances of both switches; capacitance C_{p2} represents the bottom plate parasitic capacitance of C_{R1}; capacitance C_{p3} represents the top plate parasitic capacitance of C_1, the input capacitance of the opamp, and the parasitic capacitance of switch ϕ_2; capacitance C_{p4} represents the bottom plate parasitic capacitance of C_1 and the input capacitance of the following stage.

Considering the transition $(n-1)T \rightarrow (n-0.5)T$ ($\phi_1 \rightarrow \phi_2$) and the transition $(n-0.5)T \rightarrow (n)T$ ($\phi_2 \rightarrow \phi_1$), Eq. 2.9 is obtained from adding all the capacitors that are connected to the virtual ground at the end of that transition (ϕ_2 in the first case and ϕ_1 in the second).

$$\begin{cases} [(n-1)T] \rightarrow [(n-0.5)T] : V_{in}[(n-1)T](C_{R1}+C_{p1}) - V_{out}[(n-1)T]C_1 = \\ \quad = -V_{out}[(n-0.5)T]C_1 \\ \\ [(n-0.5)T] \rightarrow [nT] : -V_{out}[(n-0.5)T]C_1 = -V_{out}[nT]C_1 \end{cases}$$

$$(2.9)$$

Combining both equations in Eq. 2.9 and using the Z-Transform, the transfer function in Eq. 2.10 is obtained.

$$H(z) = \frac{V_{out}}{V_{in}} = -\left(\frac{C_{R1}+C_{p1}}{C_1}\right)\left(\frac{z^{-1}}{1-z^{-1}}\right) \qquad (2.10)$$

Taking into the account the parasitic capacitance, it can be seen that the gain coefficient of this circuit is dependent on C_{p1}. From the calculations it can also be concluded that the parasitic capacitance C_{p3} does not influence the performance of the circuit due to the virtual ground node in the negative node of the opamp.

2.2.2 Parasitic-Insensitive Integrator

To overcome the nonlinear effect of the parasitic capacitance C_{p1}, new parasitic-insensitive structures were developed [7]. Figure 2.6 shows one of these structures.

Analyzing the charge across the capacitors in each phase, and considering that the output of the circuit will again be sampled at the end of phase ϕ_1, Table 2.3

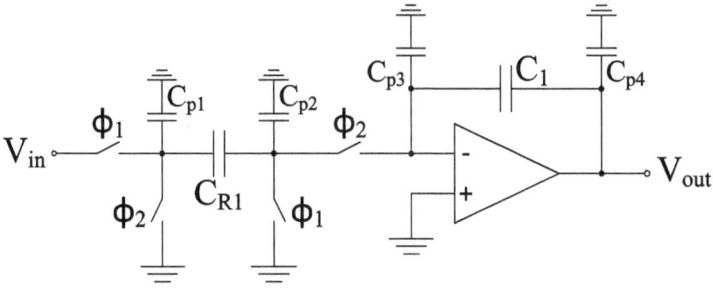

Fig. 2.6 Switched-capacitor parasitic-insensitive integrator

Table 2.3 Charge in the capacitors in each phase

	$(n{-}1)T$	$(n{-}0.5)T$	nT
$Q_{C_{R1}}$	$-V_{in}[(n-1)T]C_{R1}$	0	$-V_{in}[nT]C_{R1}$
Q_{C_1}	$-V_{out}[(n-1)T]C_1$	$-V_{out}[(n-0.5)T]C_1$	$-V_{out}[nT]C_1$
$Q_{C_{p1}}$	$V_{in}[(n-1)T]C_{p1}$	0	$V_{in}[nT]C_{p1}$
$Q_{C_{p2}}$	0	0	0
$Q_{C_{p3}}$	0	0	0
$Q_{C_{p4}}$	$V_{out}[(n-1)T]C_{p4}$	$V_{out}[(n-0.5)T]C_{p4}$	$V_{out}[nT]C_{p4}$

is obtained. Capacitance C_{p1} represents the bottom plate parasitic capacitance of C_{R1} and the parasitic capacitances of the switches connected to the bottom plate of C_{R1}; capacitance C_{p2} represents the top plate parasitic capacitance of C_{R1} and the parasitic capacitances of the switches connected to the top plate of C_{R1}; capacitance C_{p3} represents the top plate parasitic capacitance of C_1, the input capacitance of the opamp, and the parasitic capacitance of switch connected to the top plate of C_1; capacitance C_{p4} represents the bottom plate parasitic capacitance of C_1 and the input capacitance of the following stage.

Considering the transition $(n{-}1)T \rightarrow (n{-}0.5)T$ ($\phi_1 \rightarrow \phi_2$) and the transition $(n{-}0.5)T \rightarrow (n)T$ ($\phi_2 \rightarrow \phi_1$), Eq. 2.11 is obtained from adding all the capacitors that are connected to the virtual ground at the end of that transition (ϕ_2 in the first case and ϕ_1 in the second).

$$
\begin{cases}
[(n-1)T] \rightarrow [(n-0.5)T] : -V_{in}[(n-1)T]C_{R1} - V_{out}[(n-1)T]C_1 = \\
\qquad = -V_{out}[(n-0.5)T]C_1 \\
\\
[(n-0.5)T] \rightarrow [nT] : -V_{out}[(n-0.5)T]C_1 = -V_{out}[nT]C_1
\end{cases}
$$

$$(2.11)$$

Combining both equations in Eq. 2.11 and using the Z-Transform, the transfer function in Eq. 2.12 is obtained.

$$
H(z) = \frac{V_{out}}{V_{in}} = \left(\frac{C_{R1}}{C_1}\right)\left(\frac{z^{-1}}{1 - z^{-1}}\right)
\qquad (2.12)
$$

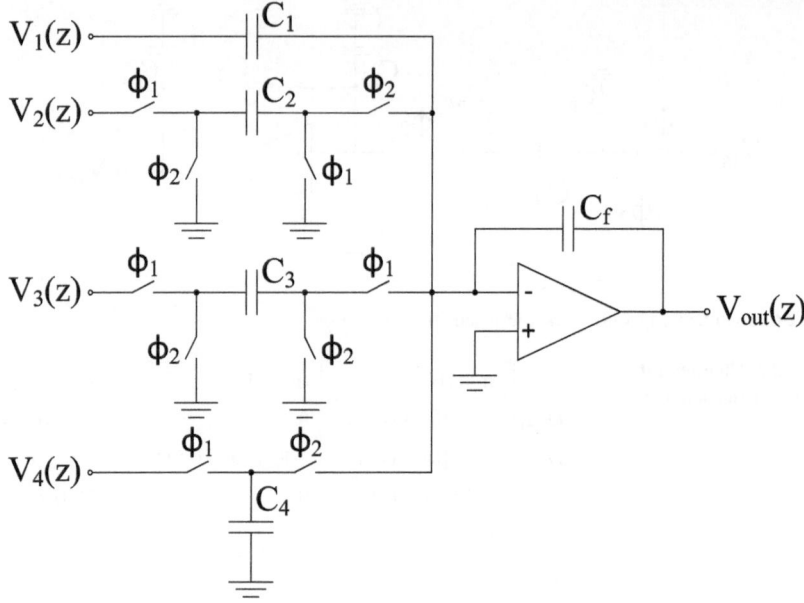

Fig. 2.7 Four-input switched-capacitor summing/integrator stage

From the calculations it can be concluded that with the addition of these two new switches the parasitic capacitances in this circuit do not influence its performance, improving the linearity of the circuit and the accuracy of the transfer function. The parasitic capacitances have little effect on the circuit since they are connected to the virtual ground (C_{p2} and C_{p3}) or the physical ground (C_{p2}). C_{p4} also has little effect since it is driven by the circuits output. Although C_{p1} is charged by the input voltage during clock phase ϕ_1, it does not affect the charge in capacitor C_{R1} since the time constant RC is long enough to charge the capacitors in less than half the clocks period, and during clock phase ϕ_2 it is discharged to ground via the ϕ_2 switch, not influencing the charge that is being transferred to C_1. Although the parasitic capacitances do not influence the circuits transfer function, they do increase the time constant RC making the settling time slower.

2.2.3 Signal Flow Graph Analysis

Using the superposition principle it is possible to determine the charge transfer functions of large circuits by analyzing the circuits graphically. Figure 2.7 shows an example of a circuit with several inputs [1].

This circuit can be analyzed using the same process as in the previous two examples. Table 2.4 shows the charge across the capacitors.

Table 2.4 Charge in the capacitors in each phase

	$(n-1)T$	$(n-0.5)T$	nT
Q_{C_f}	$-V_{out}[(n-1)T]C_f$	$-V_{out}[(n-0.5)T]C_f$	$-V_{out}[nT]C_f$
Q_{C_1}	$-V_1[(n-1)T]C_1$	$-V_1[(n-0.5)T]C_1$	$-V_1[nT]C_1$
Q_{C_2}	$-V_2[(n-1)T]C_2$	0	$-V_2[nT]C_2$
Q_{C_3}	$-V_3[(n-1)T]C_3$	0	$-V_3[nT]C_3$
Q_{C_4}	$V_4[(n-1)T]C_4$	0	$V_4[nT]C_4$

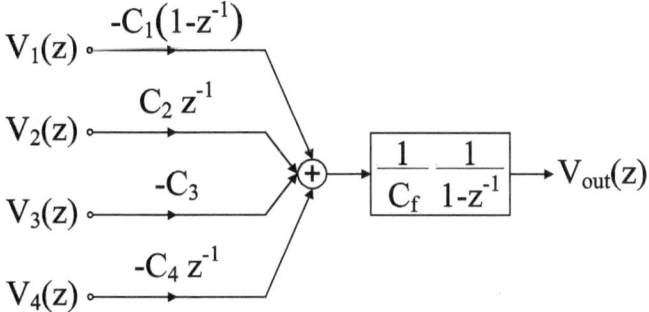

Fig. 2.8 Equivalent signal flow graph of Fig. 2.7 circuit

Analyzing the charge values in Table 2.4 individually, it is possible to obtain the charge transfer function of each subcircuit individually. These results will make it easier to analyze large circuits once the subcircuits are identified. The results of each subcircuit in Fig. 2.7 are shown in Fig. 2.8.

The signal flow graph (SFG) can be analyzed and the circuits transfer function is obtained.

$$V_{out}(z) = -\frac{C_1}{C_f}V_1(z) + \frac{C_2}{C_f}\frac{z^{-1}}{1-z^{-1}}V_2(z) - \frac{C_3}{C_f}\frac{1}{1-z^{-1}}V_3(z) - \frac{C_4}{C_f}\frac{z^{-1}}{1-z^{-1}}V_4(z)$$

(2.13)

2.3 Sallen-Key Topology

The SC low-pass and band-pass filters implemented in this book are based on the continuous time Sallen-Key second-order filters. In this section these topologies are presented and discussed.

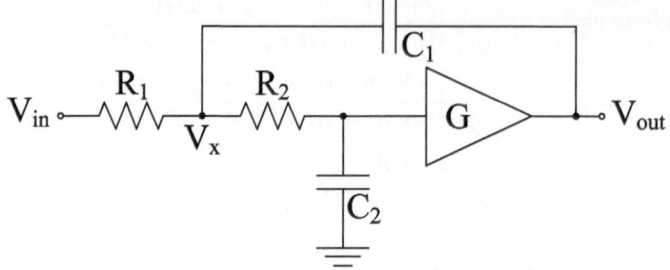

Fig. 2.9 Continuous-time Sallen-Key low-pass filter using an amplifier with gain G [8]

2.3.1 Low-Pass Sallen-Key Topology

The basis for the SC low-pass filter is the Sallen-Key low-pass filter. This filter is shown in Fig. 2.9. One of the characteristics of this type of filter is the opamp being configured as an amplifier instead of an integrator, reducing the gain-bandwidth requirements for the opamp. This means that when using this topology it is possible to implement higher frequency filters with lower gain when compared to other topologies. Another advantage of this filter is the low ratio between largest and lowest resistor/capacitor, which is important for manufacturability.

At low frequencies the capacitors can be seen as open circuits and the signal is transferred from the input to the output. At high frequency the capacitors can be seen short circuits and the signal is discharged to ground. The transfer function of this second-order filter is shown in Eq. 2.14.

$$H(s) = \frac{V_{out}}{V_{in}} = \frac{G}{1 + (C_1 R_1 + C_2 R_1 - G C_1 R_1 + C_2 R_2)s + (C_1 C_2 R_1 R_2)s^2} \tag{2.14}$$

Knowing that the transfer function of second order systems is

$$H(s) = G \frac{\omega_p^2}{s^2 + \frac{\omega_p}{Q_p}s + \omega_p^2} \tag{2.15}$$

Using Eqs. 2.14 and 2.15, the pole frequency ω_p is

$$\omega_p = \frac{1}{\sqrt{C_1 C_2 R_1 R_2}} \tag{2.16}$$

the inverse of the quality factor

$$\frac{1}{Q_p} = (1 - G)\sqrt{\frac{C_1 R_1}{C_2 R_2}} + \sqrt{\frac{C_2 R_1}{C_1 R_2}} + \sqrt{\frac{C_2 R_2}{C_1 R_1}} \tag{2.17}$$

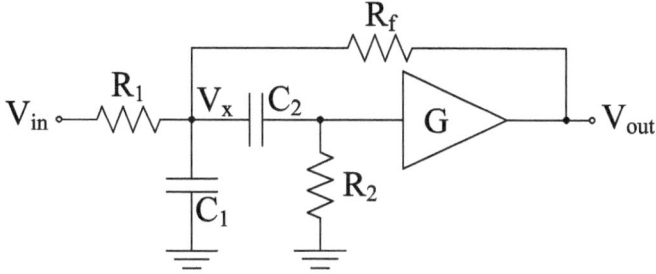

Fig. 2.10 Continuous-time Sallen-Key band-pass filter

and the low frequency gain

$$H(0) = G \tag{2.18}$$

Using these equations and selecting two of the five unknown variables it is possible to size the values of the components of the filter circuit.

2.3.2   Band-Pass Sallen-Key Topology

The basis for the SC band-pass filter is the Sallen-Key band-pass filter implemented with a voltage controlled voltage source (VCVS). The limitation of this topology is that the quality factor of the filter will determine the gain required for the amplifier. This filter is shown in Fig. 2.10.

The transfer function of this second-order filter is shown in Eq. 2.19.

$$H(s) = \frac{GC_2R_2R_fs}{R_1 + R_f + (C_1R_1R_f + C_2R_f(R_1 + R_2) - C_2R_1R_2(G-1))s + C_1C_2R_1R_2R_fs^2} \tag{2.19}$$

Where the pole frequency is

$$\omega_p = \sqrt{\frac{R_1 + R_f}{C_1C_2R_1R_2R_f}} \tag{2.20}$$

and the inverse of the quality factor

$$\frac{1}{Q_p} = \frac{C_1R_1R_f + C_2R_f(R_1 + R_2) - C_2R_1R_2(G-1)}{\sqrt{(R_1 + R_f)C_1C_2R_1R_2R_f}} \tag{2.21}$$

Chapter 3
Low-Pass Filter Topologies

Abstract Low-pass filters are systems that allow the passing without attenuation of signals with frequency below the cutoff frequency, while attenuating those with frequency above it. The amount of attenuation the signals with higher frequency suffer is dependent on their frequency and the order of the filter. In this chapter low-pass SC filters based on the continuous-time version of the Sallen-Key low-pass filter [8] will be presented and discussed when using ideal components. Higher order low-pass SC filters using cascaded sections will also be discussed in this chapter.

3.1 Continuous-Time Sallen-Key Low-Pass Filter

The continuous-time Sallen-Key low-pass filter is shown in Fig. 3.1. Applying Kirchhoff's current law (KCL), which states that the algebraic sum of currents in a network of conductors meeting at a point is zero, the equations in Eq. 3.1 are obtained.

$$\begin{cases} \dfrac{V_x - V_{in}}{R_1} + \dfrac{V_x - (1/G)V_{out}}{R_2} + \left(V_x - V_{out}\right)sC_1 = 0 \\ \dfrac{(1/G)V_{out} - V_x}{R_2} + (1/G)V_{out}sC_2 = 0 \end{cases} \tag{3.1}$$

Combining both equations (Eq. 3.1) the filters transfer function is obtained. This transfer function (Eq. 3.2) shows that the Sallen-Key topology under consideration here is a second-order low-pass filter.

$$H(s) = \frac{V_{out}}{V_{in}} = \frac{G}{1 + (C_1R_1 + C_2R_1 - GC_1R_1 + C_2R_2)s + (C_1C_2R_1R_2)s^2} \tag{3.2}$$

Several approximations can be used when designing filters. In this case the filter will be designed using the Butterworth prototype transfer function which is described in Appendix A.1.

© Springer International Publishing Switzerland 2015

H. A. de A. Serra, N. Paulino, *Design of Switched-Capacitor Filter Circuits using Low Gain Amplifiers,* SpringerBriefs in Electrical and Computer Engineering, DOI 10.1007/978-3-319-11791-1_3

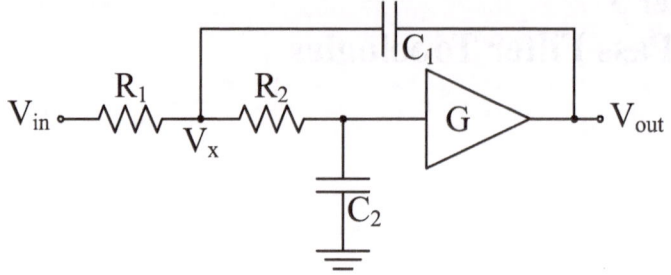

Fig. 3.1 Continuous-time Sallen-Key low-pass filter

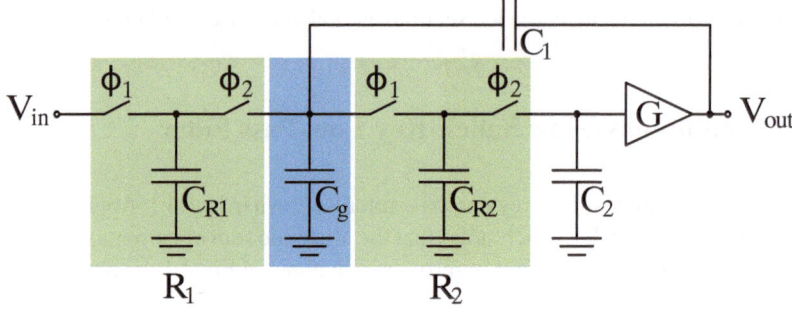

Fig. 3.2 Single-ended low-pass SC filter [9]

3.2   Switched-Capacitor Low-Pass Filter

3.2.1   Single-Ended Switched-Capacitor Low-Pass Filter

The single-ended version of the SC low-pass filter is shown in Fig. 3.2. This circuit is based on the Sallen-Key low-pass filter (Fig. 3.1) replacing the resistors with parallel SC networks and adding a capacitor C_g connected to ground in node V_x. This capacitor was added since there is a small parasitic capacitance in node V_x that causes large variations in the voltage of the node, preventing the circuit from working properly. By adding C_g with significant capacitance ($\simeq 500$ fF), the voltage in the node will be well defined.

Unlike analog filters, which are analyzed using Kirchhoff's voltage law (KVL) and KCL, SC circuits are analyzed from a charge conservation perspective. Figure 3.3a, b show the equivalent circuit during clock phase ϕ_1 and clock phase ϕ_2, respectively.

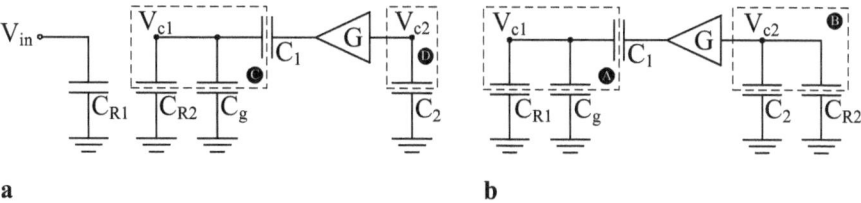

Fig. 3.3 Low-pass SC filter during: (**a**) Clock phase 1 (**b**) Clock phase 2

Assuming charge conservation between phases and considering that the signal is sampled at the end of phase 1, the charge equations are obtained.

$$
\begin{cases}
A \longrightarrow V_{in}[n-1]C_{R1} + V_{c1}[n-1](C_1 + C_g) - GV_{c2}[n-1]C_1 = \\
\qquad = V_{c1}[n-0.5](C_{R1} + C_1 + C_g) - GV_{c2}[n-0.5]C_1 \\
B \longrightarrow V_{c1}[n-1]C_{R2} + V_{c2}[n-1]C_2 = V_{c2}[n-0.5](C_2 + C_{R2}) \\
C \longrightarrow V_{c2}[n-0.5](C_{R2} - GC_1) + V_{c1}[n-0.5](C_g + C_1) = \\
\qquad = V_{c1}[n](C_{R2} + C_1 + C_g) - GV_{c2}[n]C_1 \\
D \longrightarrow V_{c2}[n-0.5]C_2 = V_{c2}[n]C_2
\end{cases}
\tag{3.3}
$$

From the equations in Eq. 3.3 and knowing that $V_{c2} = V_{out}/G$, the filters transfer function is obtained. This transfer function (Eq. 3.4) shows that the obtained SC filter is also a second-order low-pass filter.

$$
H(z) = \frac{V_{out}}{V_{in}} = G\frac{d}{a - bz + cz^2}
\tag{3.4}
$$

where the coefficients a, b, c, and d are given by:

$$
\begin{cases}
a = & C_1^2 C_2 + 2C_1 C_2 C_g + C_2 C_g^2 + GC_1^2 C_{R2} + GC_1 C_g C_{R2} \\
b = & 2C_1^2 C_2 + 4C_1 C_2 C_g + 2C_2 C_g^2 + C_1 C_2 C_{R1} + C_2 C_g C_{R1} + C_1^2 C_{R2} + \\
& + C_1 C_2 C_{R2} + 2C_1 C_g C_{R2} + C_2 C_g C_{R2} + C_g^2 C_{R2} + C_2 C_{R1} C_{R2} + \\
& + C_1 C_{R2}^2 + C_g C_{R2}^2 + C_{R1} C_{R2}^2 + GC_1^2 C_{R2} + GC_1 C_g C_{R2} \\
c = & C_1^2 C_2 + 2C_1 C_2 C_g + C_2 C_g^2 + C_1 C_2 C_{R1} + C_2 C_g C_{R1} + C_1^2 C_{R2} + \\
& + C_1 C_2 C_{R2} + 2C_1 C_g C_{R2} + C_2 C_g C_{R2} + C_g^2 C_{R2} + C_1 C_{R1} C_{R2} + \\
& + C_2 C_{R1} C_{R2} + C_g C_{R1} C_{R2} + C_1 C_{R2}^2 + C_g C_{R2}^2 + C_{R1} C_{R2}^2 \\
d = & (C_1 + C_g)C_{R1} C_{R2}
\end{cases}
\tag{3.5}
$$

As for analog filters, the coefficients of the discrete-time transfer function can be obtained using different methods. These methods are described in Appendix A.2. Specifying a cutoff frequency of 1 MHz, a quality factor Q_p of $1/\sqrt{2}$, and a clock

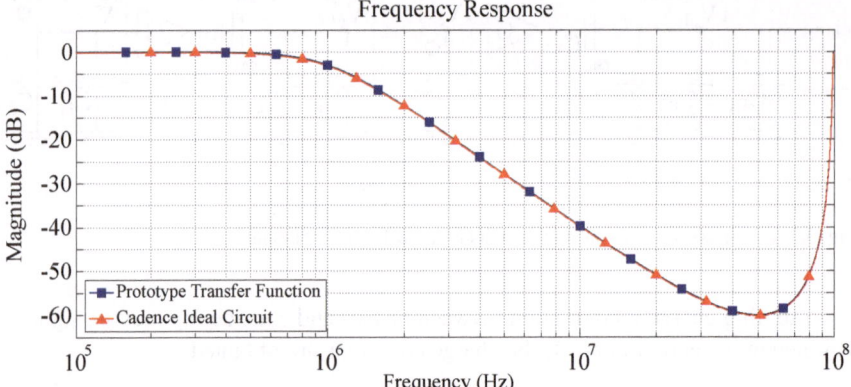

Fig. 3.4 Discrete-time frequency responses

frequency of 100 MHz, the coefficients of the transfer function are obtained. Considering that the filter is implemented using a buffer with a gain of 0.98 and $C_1 = 4$ pF, $C_{R2} = 140$ fF, and $C_g = 400$ fF, the values obtained are $C_{R1} \simeq 218.7$ fF and $C_2 \simeq 1.56$ pF.

The single-ended version of the SC filter (Fig. 3.2) was simulated in Cadence using ideal components (switches, buffer, and capacitors) and capacitors with the previously mentioned values. To determine if the circuit is working correctly, the impulse response of the filter was simulated, and used, to plot the filters Bode diagram. The process to obtain the impulse response and the Bode diagram is described in Appendix B.

Figure 3.4 shows the frequency response of prototype transfer function and the response from simulating the filter with ideal components in single-ended configuration. Since no solution was found for the capacitors of this filter, when using the numerator and denominator of the prototype transfer function, the filter was sized using only the denominator of the transfer function.

Using only the denominator will determine the DC gain of the filter as the gain of the buffer used in the filter and thus why the ideal response of the filter is slightly shifted down 0.18 dB from the prototype response. The attenuations at the cutoff frequency were 3 dB for the prototype transfer function and 3.18 dB for the filter using ideal components.

3.2.2   Differential Switched-Capacitor Low-Pass Filter

In order to transform the filter from the single-ended to the differential configuration, the relation between the capacitors in these two configurations must be calculated. Figure 3.5 shows the difference between the two.

Fig. 3.5 Capacitor in
single-ended (**a**) and
differential configuration (**b**)

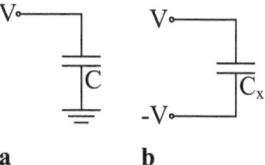

Since the charge across the capacitor C is equal to $C(V-0)$ and the charge across C_x is equal to $C_x(V-(-V))$, this means that the capacitor C_x must have half the capacitance of C in order to store the same charge as capacitor C, assuming that V has the same value in both cases. The differential configuration is shown in Fig. 3.6.

The differential configuration of the SC filter was simulated and the frequency response was compared with the single-ended configuration (Fig. 3.7). This simulation shows that the relation between capacitance values in single-ended and differential configuration is correct.

3.3   Switched-Capacitor Filters Using Cascaded Sections

Higher order SC filters can be obtained by cascading lower order filters. In this section a sixth-order low-pass filter will be built using three cascaded second-order sections, maintaining a 3 dB attenuation at the cutoff frequency.

In order to obtain this attenuation at the cutoff frequency, each of the second-order sections must have a specific quality factor. Using Eq. A.2 with $n = 6$, the denominators for the three second-order sections (Eq. 3.6) can be obtained.

$$T(\hat{S}) = \frac{1}{\hat{S}^2 + 0.5176\hat{S} + 1} \frac{1}{\hat{S}^2 + 1.4142\hat{S} + 1} \frac{1}{\hat{S}^2 + 1.9319\hat{S} + 1} \qquad (3.6)$$

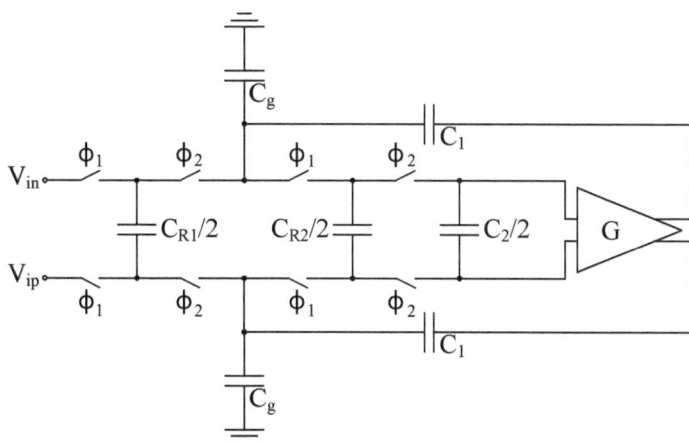

Fig. 3.6 Differential low-pass SC filter

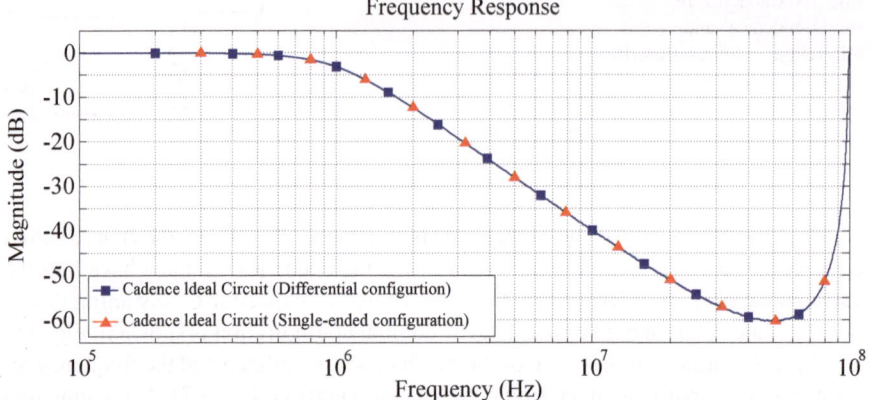

Fig. 3.7 Frequency response comparison between single-ended and differential configurations

Table 3.1 Transfer function coefficients obtained from Matlab

Q_p	Poles		Numerator	Denominator		
	A	$\pm j$B	z^0	z^2	z^1	z^0
1.93185	0.98207	0.05968	3.8e-03	1	-1.9641	0.96802
0.70711	0.95560	0.04251	3.8e-03	1	-1.9112	0.91498
0.51764	0.94096	0.01532	3.8e-03	1	-1.8819	0.88563

Table 3.2 Capacitance values for each section

Q_p	Gain	C_1	C_2	C_{R1}	C_{R2}	C_g
1.93185	0.98	16 pF	444.658 fF	350.104 fF	100 fF	100 fF
0.70711	0.98	4 pF	1.56034 pF	218.717 fF	140 fF	400 fF
0.51764	0.98	4 pF	2.83061 pF	390.231 fF	140 fF	400 fF

Using these coefficients the quality factors necessary for each section can be obtained in order to have the desired 3 dB attenuation and with Eq. A.9 the coefficients for each section can be obtained. Alternatively Matlab can be used for higher order filters (Appendix A.2). Using this approach the coefficients in Table 3.1 are obtained.

Using the denominator coefficients shown in Table 3.1, the capacitance values were calculated for each section. The values obtained for the capacitors are shown in Table 3.2.

The frequency response obtained for each of the sections is shown in Fig. 3.8. The gain at the cutoff frequency for Q_p equal to 0.5176 was -5.71 dB for the prototype transfer function and -5.89 dB for the filter using ideal components. For Q_p equal to 0.7071 was -3.00 and -3.18 dB, respectively. For Q_p equal to 1.9319 was 5.73 and 5.55 dB, respectively.

To cascade the second-order filters into a sixth-order filter the output of the filters are connected to the input of the following section. The obtained filter response is

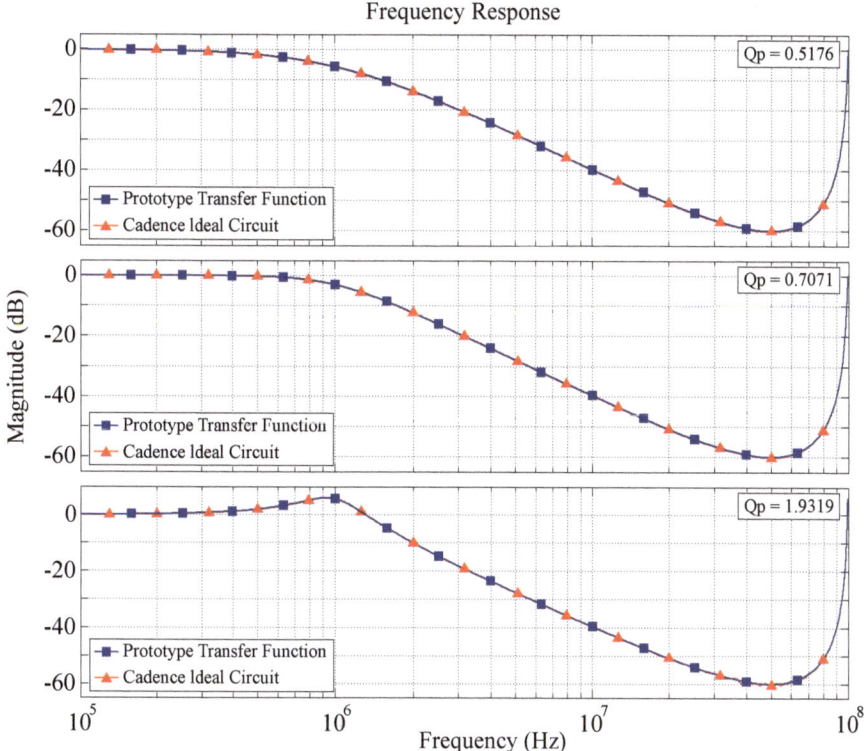

Fig. 3.8 Frequency response of each second-order section

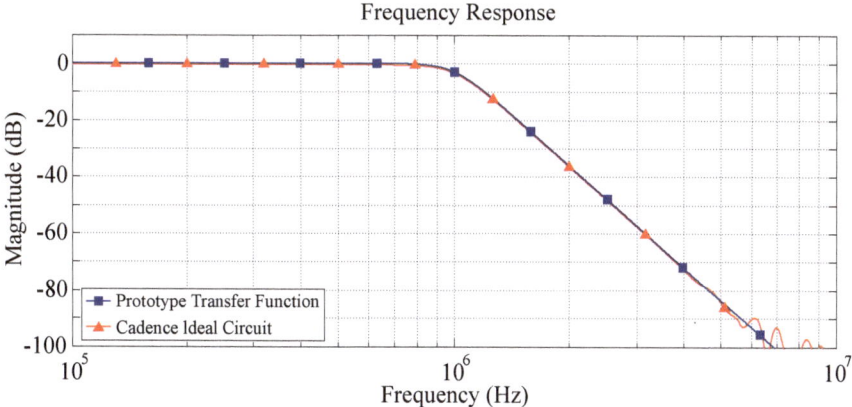

Fig. 3.9 Frequency response of the sixth-order filter

independent of the order in which the sections are connected. The obtained frequency response for the sixth-order filter is shown in Fig. 3.9. At the cutoff frequency the prototype transfer function has an attenuation of 2.99 dB and the filter using ideal components has an attenuation of 3.51 dB. The difference in attenuations is due to the gain of the buffers, since each section loses approximately 0.18 dB due to the buffer, shifting down a total of 0.53 dB from the prototype transfer function.

3.4   Conclusions

This chapter presented, from an ideal point of view, that it is possible to implement low-pass SC filters based on the Sallen-Key low-pass filter using near unity-gain buffers. It was also described how to obtain higher order filters based on second-order sections while maintaining an attenuation of approximately -3 dB at the cutoff frequency. The differences between the prototype transfer function frequency response and the ideal circuits response was due to the buffers gain. Since the filter was designed considering only the denominator of the prototype transfer function, the DC gain of the filter is equal to the gain of the buffer.

Chapter 4
Band-Pass Filter Topologies

Abstract Band-pass filters are systems that allow the passing of signals within a band of frequencies, while attenuating the frequencies outside the defined ranged. The amount of attenuation the signals suffers outside the frequency range is dependent on the signals frequency and the order of the filter. In this chapter band-pass SC filters based on the continuous-time version of the Sallen-Key band-pass filter, implemented with a VCVS, will be presented and discussed when using ideal components. Higher order band-pass SC filters using cascaded sections will also be discussed in this chapter.

4.1 Continuous-time Sallen-Key Band-Pass Filter

The continuous-time Sallen-Key band-pass filter is shown in Fig. 4.1. Applying KCL to the circuit, the equations in Eq. 4.1 are obtained.

$$\begin{cases} \dfrac{V_x - V_{in}}{R_1} + \dfrac{V_x - V_{out}}{R_f} + (V_x - (1/G)V_{out})sC_2 + V_x sC_1 = 0 \\[2mm] ((1/G)V_{out} - V_x)sC_2 + \dfrac{(1/G)V_{out}}{R_2} = 0 \end{cases} \tag{4.1}$$

Combining both equations (Eq. 4.1) the filters transfer function is obtained. This transfer function (Eq. 4.2) shows that the Sallen-Key topology under consideration here is a second-order band-pass filter (one zero and two poles).

$$H(s) = \frac{GC_2 R_2 R_f s}{R_f + C_2 R_2 R_f s + R_1(1 + C_1 R_f s + C_2 s(R_2 - GR_2 + R_f + C_1 R_2 R_f s))} \tag{4.2}$$

Two methods to design a continuous-time band-pass filter using the Butterworth prototype transfer function are described in Appendix A.3.

© Springer International Publishing Switzerland 2015

H. A. de A. Serra, N. Paulino, *Design of Switched-Capacitor Filter Circuits using Low Gain Amplifiers,* SpringerBriefs in Electrical and Computer Engineering, DOI 10.1007/978-3-319-11791-1_4

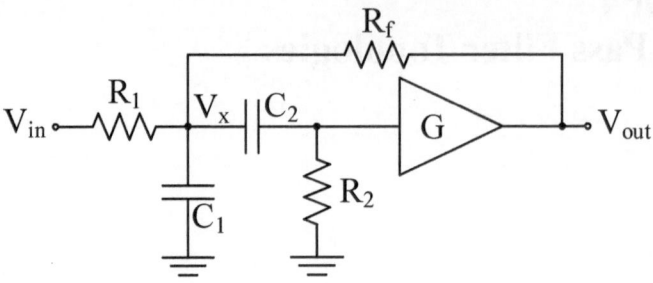

Fig. 4.1 Continuous-time Sallen-Key band-pass filter

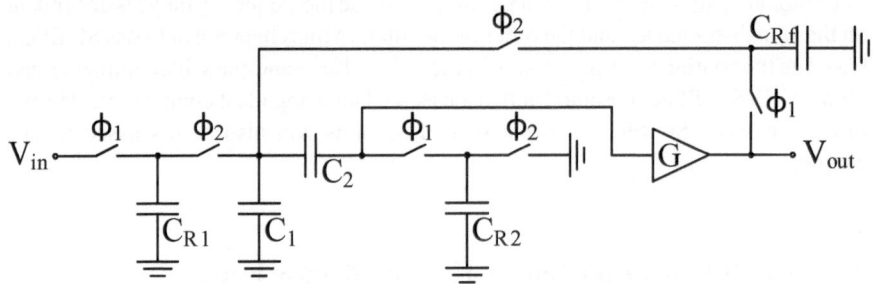

Fig. 4.2 Single-ended band-pass SC filter

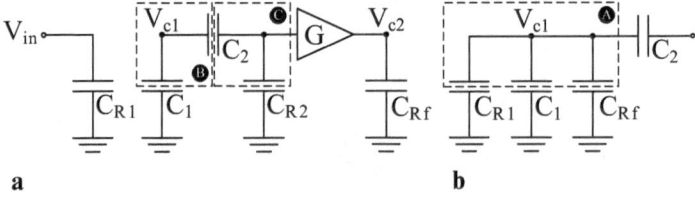

Fig. 4.3 Band-pass SC filter during: **a** Clock phase 1 **b** Clock phase 2

4.2 Switched-Capacitor Band-Pass Filter

4.2.1 Single-Ended Switched-Capacitor Band-Pass Filter

The single-ended version of the SC band-pass filter is shown in Fig. 4.2. This circuit is obtained from the continuous-time version of the filter shown in the previous section (Sect. 4.1), by replacing the resistors with parallel SC networks.

Figure 4.3a and b show the equivalent circuit during clock phase ϕ_1 and clock phase ϕ_2, respectively.

Assuming charge conservation between phases and considering that the signal is sampled at the end of phase 1, the charge equations (Eq. 4.3) are obtained. During clock phase 2 the capacitor C_2 is floating, meaning that there is no current flowing through it and therefore this capacitor maintains the charge that was stored during phase 1. This fact must be taken into account when extracting the charge conservation equations between phases.

$$\begin{cases} A \longrightarrow V_{in}[n-1]C_{R1} + V_{c1}[n-1]C_1 + GV_{c2}[n-1]C_{Rf} = \\ \qquad = V_{c1}[n-0.5](C_{R1} + C_1 + C_{Rf}) \\ B \longrightarrow V_{c1}[n-1]C_2 + V_{c1}[n-0.5]C_1 - V_{c2}[n-1]C_2 = \\ \qquad = V_{c1}[n](C_1 + C_2) - V_{c2}[n]C_2 \\ C \longrightarrow V_{c2}[n-1]C_2 - V_{c1}[n-1]C_2 = V_{c2}[n](C_{R2} + C_2) - V_{c1}[n]C_2 \end{cases}$$ (4.3)

From the equations in Eq. 4.3 and knowing that $V_{c2} = V_{out}/G$, the filters transfer function is obtained. This transfer function (Eq. 4.4) shows that the obtained SC filter is also a second-order band-pass filter.

$$H(z) = \frac{V_{out}}{V_{in}} = G\frac{d(z-1)}{a - bz + cz^2}$$ (4.4)

where the coefficients a, b, c, and d are given by:

$$\begin{cases} a = C_1^2 C_2 + GC_1 C_2 C_{Rf} \\ b = 2C_1^2 C_2 + C_1 C_2 C_{R1} + C_1^2 C_{R2} + C_1 C_2 C_{R2} + C_2 C_{R1} C_{R2} + C_1 C_2 C_{Rf} + \\ \qquad + C_2 C_{R2} C_{Rf} + GC_1 C_2 C_{Rf} \\ c = C_1^2 C_2 + C_1 C_2 C_{R1} + C_1^2 C_{R2} + C_1 C_2 C_{R2} + C_1 C_{R1} C_{R2} + C_2 C_{R1} C_{R2} + \\ \qquad + C_1 C_2 C_{Rf} + C_1 C_{R2} C_{Rf} + C_2 C_{R2} C_{Rf} \\ d = C_1 C_2 C_{R1} \end{cases}$$

(4.5)

Two methods to design a discrete-time band-pass filter using the Butterworth prototype transfer function are described in Appendix A.3. Specifying a central frequency of 1 MHz, a quality factor $Q_p = 2$ (pass band of 500 kHz), and a clock frequency of 100 MHz, the lateral frequencies f_L and f_H are equal to 780.8 and 1280.7 kHz, respectively. Considering that the filter is implemented using an amplifier with gain of 1.25 and capacitance $C_1 = 3$ pF and $C_2 = 4$ pF, the values obtained are $C_{R1} \simeq 99.5$ fF, $C_{R2} \simeq 80.7$ fF, and $C_{Rf} \simeq 673.4$ fF.

The single-ended version of the SC filter (Fig. 4.2) was implemented using ideal components (switches, buffer, and capacitors) and capacitors with the previously mentioned values. The filters frequency responses are shown in Fig. 4.4. At the cutoff frequencies the simulated filter circuit has an attenuation of [2.57, 2.27] dB and [2.15, 1.88] dB, and apart from a small gain difference (approximately 0.4 dB) the frequency responses are identical.

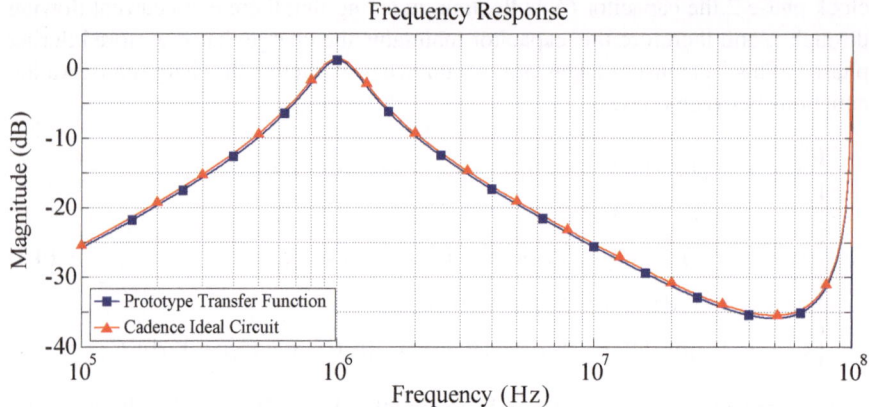

Fig. 4.4 Discrete-time frequency responses

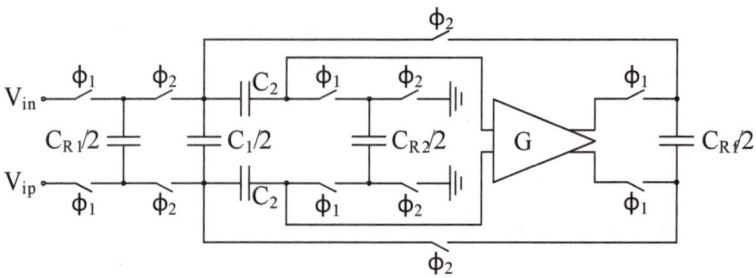

Fig. 4.5 Differential band-pass SC filter

4.2.2 Differential Switched-Capacitor Band-Pass Filter

The differential configuration is shown in Fig. 4.5 and was obtained using the same method that was previously described in Sect. 3.2.2.

The differential configuration of the SC filter was simulated and the frequency response was compared with that of the single-ended configuration (Fig. 4.6). This simulation shows that the relation between capacitance values in single-ended and differential configuration is correct.

4.3 Switched-Capacitor Filters Using Cascaded Sections

Higher order SC filters can be obtained by cascading lower order filters. In this section a fourth-order band-pass filter will be built using two cascaded second-order sections with the same quality factor ($Q_p = 1.285$).

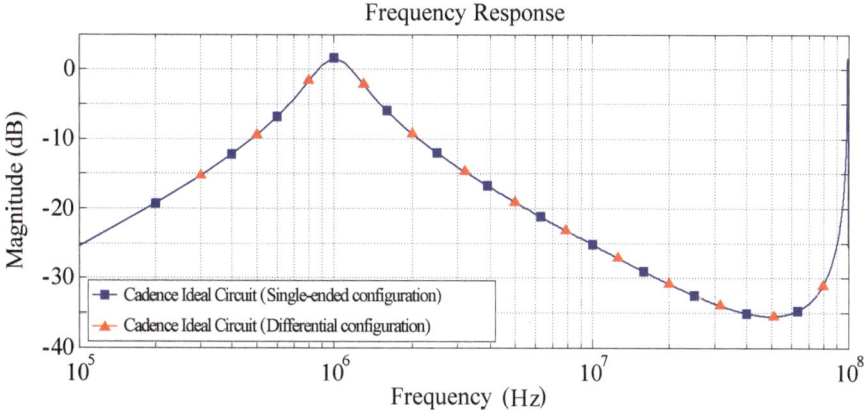

Fig. 4.6 Frequency response comparison between single-ended and differential configurations

Table 4.1 Transfer function coefficients obtained from the prototype method

Q_p	Poles		Numerator		Denominator		
	A	$\pm j$B	z^1	z^0	z^2	z^1	z^0
1.285	0.97555	0.05788	48.8964e-03	-48.8964e-03	1	-1.9511	0.95501

Table 4.2 Capacitance values for the cascaded sections

Q_p	Gain	C_1	C_2	C_{R1}	C_{R2}	C_{Rf}
1.285	1.25	3 pF	4 pF	154.535 fF	80.7119 fF	618.398 fF

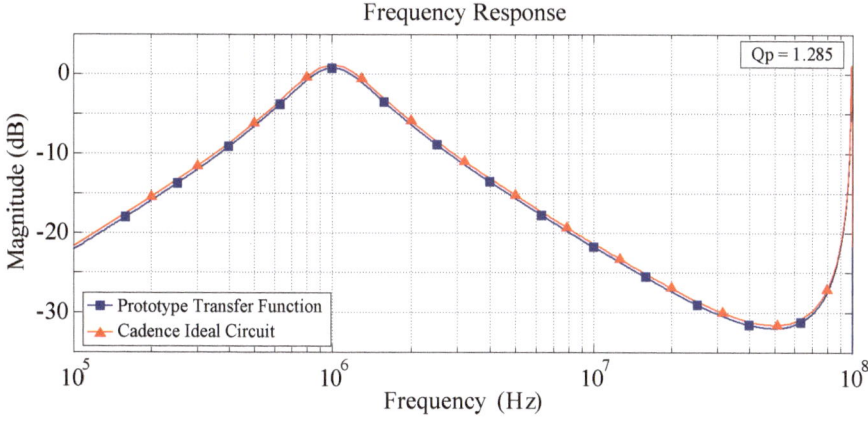

Fig. 4.7 Frequency response of the second-order section

The coefficients for the cascaded second-order sections can be obtained using Eq. A.15 with $Q_p = 1.285$. This quality factor will define a 1.5 dB attenuation at the cutoff frequencies for each section and 3 dB for the cascaded section. Table 4.1 shows the poles location and the transfer function coefficients.

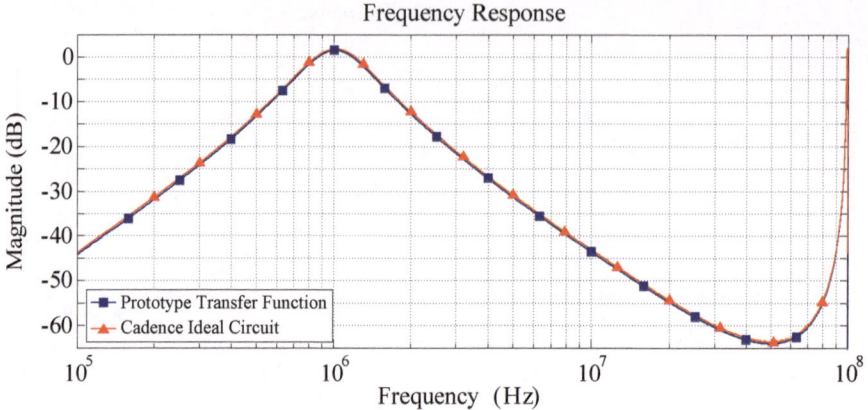

Fig. 4.8 Frequency response of the fourth-order filter

Using the denominator coefficients shown in Table 4.1, the capacitance values were calculated for each section. The obtained values for the capacitors are shown in Table 4.2.

The frequency response obtained for this section is shown in Fig. 4.7. Cascading the two second-order filters the frequency response obtained for the fourth-order filter is shown in Fig. 4.8.

4.4   Conclusions

This chapter presented, from an ideal point of view, that it is possible to implement band-pass SC filters based on the Sallen-Key band-pass filter using low-gain amplifiers. It was also described how to obtain higher order filters based on second-order sections. Apart from a small gain difference the frequency responses of the prototype transfer function and of the ideal circuit are identical.

Chapter 5
Non-Ideal Effects

Abstract In the third and fourth chapters, the topologies of a low-pass and a band-pass filter have been presented using ideal components. In this chapter the non-ideal effects from using real components will be presented and discussed. Initially, the non-ideal effects due to the switches are described. The distortion of a first-order low-pass SC filter is simulated using different switch configurations, including with the use of clock boost circuits, in order to determine which configuration is better from a distortion point of view. Two clock boost circuits, one of them capable of operating at lower voltages, are presented. A voltage buffer, and a low gain amplifier are described in several configurations; the low and medium frequency small signal models are presented and the respective transfer functions derived; the thermal noise referred to the input is also derived. The chapter ends with some simulation results, showing power dissipation, gain, gain-bandwidth product, THD, and thermal noise for each version of the circuits.

5.1 Non-linear Effects due to Real Switches

Ideal switches have a very small constant resistance when they are closed, and a very high constant resistance when they are open. Real switches (transistors), however, have considerable resistance when they are closed and the value depends on the V_{GS} voltage of the transistor, introducing distortion in the signal. Another source of distortion is the parasitic capacitances of the real switch since their value also changes with the four node voltages. In addition to the variation with the voltage, both the resistance and the parasitic capacitances change with the switches size. To determine the relation between the non-linear effects of the switches and their sizes, a first-order low-pass SC filter will be analyzed and simulated, since it's a simple circuit with only two switches in single-ended configuration.

5.1.1 Filter Analysis

Figure 5.1 shows a passive first-order low-pass filter from which the SC version is derived.

© Springer International Publishing Switzerland 2015

H. A. de A. Serra, N. Paulino, *Design of Switched-Capacitor Filter Circuits using Low Gain Amplifiers*, SpringerBriefs in Electrical and Computer Engineering, DOI 10.1007/978-3-319-11791-1_5

Fig. 5.1 Passive first-order
low-pass filter

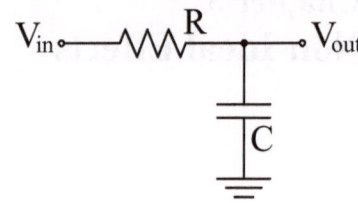

Fig. 5.2 First-order low-pass
SC filter

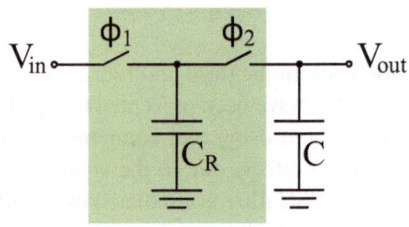

The SC version of the filter, which is shown in Fig. 5.2, is obtained by replacing the resistor R with a parallel SC network.

Assuming charge conservation between phases and starting the charge analysis from phase 1 to phase 2, the charge equations are obtained.

$$\begin{cases} V_{in}[n-1]C_R + V_{out}[n-1]C = V_{out}[n-0.5](C_R+C) \\ V_{out}[n-0.5]C = V_{out}[n]C \end{cases} \tag{5.1}$$

From these equations the first-order low-pass transfer function (Eq. 5.2) is obtained.

$$H(z) = \frac{V_{out}}{V_{in}} = \frac{C_R}{(C_R+C)z - C} \tag{5.2}$$

The coefficients C_R and C can be obtained using one of the methods described in Appendix A.2 for a first-order filter.

5.1.2 Simulation Results

Specifying a cutoff frequency of 1 MHz, a clock frequency of 100 MHz, and the capacitor $C = 2.5$ pF, the value obtained for C_R is approximately 162.2 fF. The frequency response using these specifications is shown in Fig. 5.3.

To determine the influence of the non-linear effects of the resistance $(1/g_{ds})$ and the parasitic capacitances of the switches, the distortion of the filter was simulated using three versions of the SC filter (in differential configuration). In one version, the filter was simulated using ideal switches. This simulation will give the best obtainable distortion since it is not affected by the non-linear effects of switches and buffer. In the second version, the filter was simulated using a real switch. The

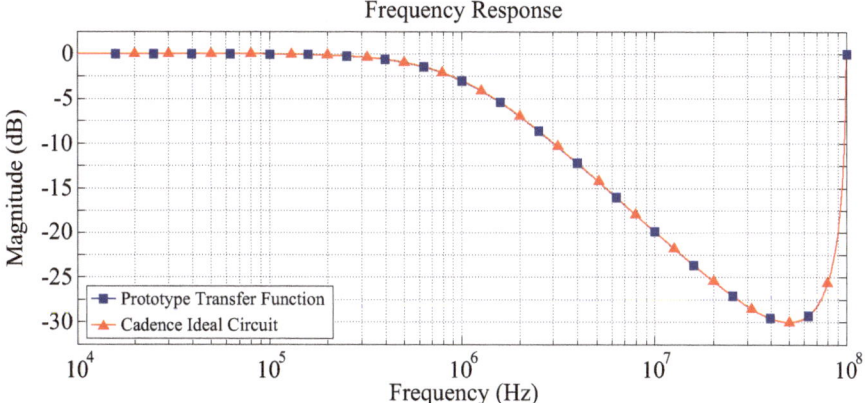

Fig. 5.3 First-order filter frequency response

Table 5.1 First-order filter distortions: (1) NMOS switch driven by a 1.2 V clock signal, (2) NMOS switch driven by a clock boosted signal, (3) Transmission gate driven by 1.2 V clock signal

		THD (dB)				
Version	Type of switch	$W = 0.16\ \mu m$	$W = 1\ \mu m$	$W = 3\ \mu m$	$W = 7\ \mu m$	$W = 10\ \mu m$
1	Ideal	−105.02	−105.02	−102.90	−102.90	−103.01
	Real	−15.70	−14.12	−15.53	−17.51	−18.69
	Parallel	−100.72	−96.70	−88.58	−82.39	−78.60
2	Ideal	−104.27	−104.27	−105.51	−105.51	−105.21
	Real	−82.31	−64.20	−55.19	−50.28	−48.76
	Parallel	−100.26	−100.46	−76.41	−69.59	−70.19
3	Ideal	−103.13	−103.60	−103.54	−103.82	−103.43
	Real	−60.25	−71.13	−64.44	−61.97	−60.89
	Parallel	−99.87	−92.27	−83.16	−75.88	−73.28

distortion obtained from this simulation is affected by the non-linear effects of both resistance and parasitic capacitances. In the third version, the filter was simulated using a real switch in parallel with an ideal switch. By doing this the non-linear effects due to the switches resistance are eliminated since the equivalent resistance of the switch is determined by the 1 Ω resistance of the ideal switch. The distortion obtained from this simulation is then only affected by the non-linear effects of the parasitic capacitances.

The simulation results are shown in Table 5.1 for three different switches, with a sinusoidal differential signal with 600 mV of amplitude. In the transmission gate switch, the PMOS transistor is four times the size of the NMOS transistor to approximate the values of their resistance. All switches were simulated using minimum length (L = 120 nm).

Table 5.2 Influence of the resistance and parasitic capacitances in the overall distortion: (1) NMOS switch driven by a 1.2 V clock signal, (2) NMOS switch driven by a clock boosted signal, (3) Transmission gate driven by 1.2 V clock signal

Influence due to	Version	Distortion (dB)				
		$W = 0.16\,\mu m$	$W = 1\,\mu m$	$W = 3\,\mu m$	$W = 7\,\mu m$	$W = 10\,\mu m$
Parasitic capacitances	1	4.3086	8.3282	14.3186	20.5033	24.4119
	2	4.0116	3.8129	29.0997	35.9156	35.0181
	3	3.2598	11.3249	20.3846	27.9446	30.1463
Parasitic capacitances and resistance	1	89.3278	90.9076	87.3650	85.3835	84.3211
	2	21.9645	40.0682	50.3245	55.2281	56.4415
	3	42.8830	32.4657	39.0956	41.8570	42.5393
Resistance	1	85.0192	82.5794	73.0464	64.8802	59.9092
	2	17.9529	36.2553	21.2247	19.3125	21.4235
	3	39.6232	21.1408	18.7110	13.9124	12.3931

From Table 5.1 it can be concluded that when driven by a 1.2 V clock signal, the NMOS switch has very high distortion. The source of this distortion is the switches resistance that is very non-linear with the variation of the input voltage. In this case it can be seen that the non-linear resistance dominates over the non-linear parasitic capacitances. When driven by a clock boosted signal, the distortion introduced by the NMOS switch decreases considerably. This is due to the decrease in the non-linear variation of the switches resistance with the input voltage, which will be seen later. The distortion introduced by a transmission gate driven by a 1.2 V clock signal also decreases, but less than in the clock boosted NMOS switch.

Table 5.2 quantifies how much the non-linear effects of the MOS switches, i. e., the non-linear resistance and the non-linear parasitic capacitances, increase the distortion of the SC circuit. The objective of this table is to clarify which non-linear effect dominates the overall distortion of the circuit when the width of the switches varies. The values presented in the table are calculated based on the distortion values presented on Table 5.1 and represent the added distortion that each non-linear effect of the switch causes in the circuit when compared to using ideal switches.

The added distortion of the parasitic capacitances is calculated subtracting the distortion of the ideal circuit from the distortion of the circuit with ideal switches in parallel with real switches. The added distortion of the real switches (including both non-linear resistance and non-linear parasitic capacitances) is calculated subtracting the distortion of the ideal circuit from the distortion of the real circuit. Finally in order to better understand the dominant cause of distortion in the circuit, the difference between the distortion of the real circuit and the distortion of the circuit with the ideal switches in parallel with the real switches is also calculated.

From Table 5.2 it can be seen that the distortion introduced by the parasitic capacitances increases with the size of the switch, which means that the non-linearity of these capacitances increases with the width of the transistor and will become harder to compensate their effect in larger switches. As for the switches resistance,

Fig. 5.4 Ideal charge pump
representation

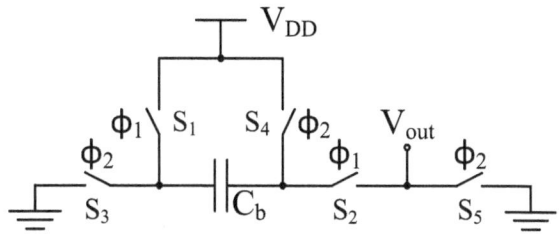

its non-linearity decreases with the switches width causing a decrease in the distortions value. When using a clock boosted signal the simulation results show that the
resistance variation with the switches width is no longer linear.

Comparing the results it can be concluded that using a clock boosted signal to
drive the NMOS switch provides lower distortion when compared with the other
two versions and the lowest width possible should be used while maintaining a time
constant small enough to charge the capacitors. The clock boost circuit used in these
simulations will be discussed in Sect. 5.2, along with a more detailed analysis of
what happens to the switches resistance and parasitic capacitances when using a
clock boosted signal to drive the switches.

5.2   Clock Boost Circuit

In this section the clock boost circuit used in Sect. 5.1 is analyzed and simulated.
This circuit is used to generate a voltage level higher than the chips supply voltage
from which it operates, and is then used to drive the NMOS switches in the SC filter.

5.2.1   Clock Boost Circuit Analysis

An ideal representation of a clock boost circuit (charge pump) is shown in Fig. 5.4.

During clock phase ϕ_1 the capacitor C_b is charged to the supply voltage V_{DD}.
During clock phase ϕ_2 one of the capacitors plates assumes a potential of V_{DD},
while the capacitor maintains the charge $Q = C_b \times V_{DD}$ from the previous clock
phase. This means, assuming charge conservation between phases, that the charge
equation is given by Eq. 5.3.

$$Q = C_b(V_{out} - V_{DD}) = C_b(V_{DD} - 0) \tag{5.3}$$

Solving Eq. 5.3 in order of V_{out}, the charge pump has an output voltage of $2 \times V_{DD}$.
The practical realization of the charge pump is shown in Fig. 5.5. This circuit, using
clock phases ϕ and ϕ_n, generates a boosted phase ϕ_B, that is a boosted version of
clock phase ϕ.

The signal V_{in} shown in Fig. 5.5 can either be the voltage at the entrance of the
switch or the supply voltage (V_{DD}). In the first case, switch S_1 shown in Fig. 5.4

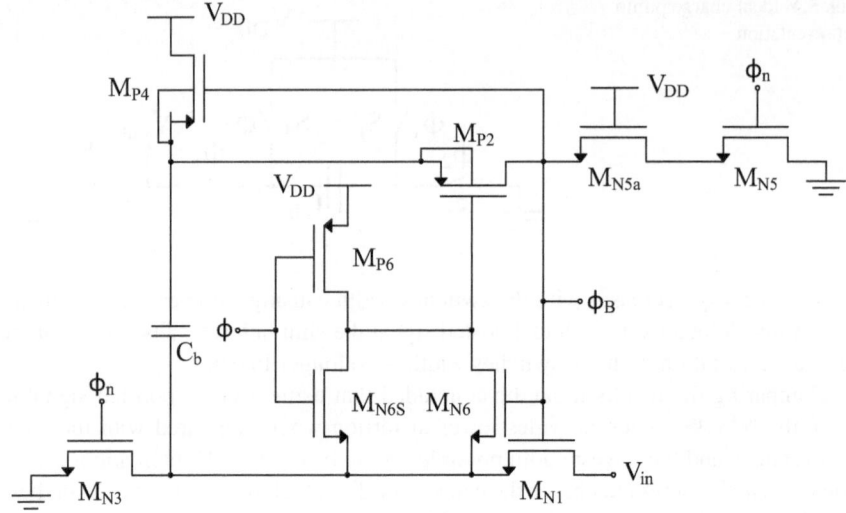

Fig. 5.5 Clock boost circuit [10]

would be connected to V_{in} instead of V_{DD} and the output voltage would be given
by Eq. 5.4. By using this configuration the output voltage of the clock boost circuit
will change depending on its input, meaning that ideally the V_{GS} voltage will remain
constant. Since the switches resistance $(1/g_{ds})$ varies with V_{GS}, in this case the
distortion due to the non-linear resistance of the switches can be greatly reduced.
One of the problems with this approach is that the filters transfer function would
have to be recalculated, since the input capacitance of each clock boost circuit will
be connected at the entrance of each switch.

$$Q = C_b(V_{out} - V_{in}) = C_b(V_{DD} - 0) \Rightarrow V_{out} = V_{DD} + V_{in} \qquad (5.4)$$

An alternative method, although not has good as the previous one in terms of
distortion, is to have a constant V_{DD} voltage at the input of the clock boost circuits
and this way the filters transfer function does not have to recalculated. Using this
configuration the output of the clock boost circuit will be, ideally, twice the supply
voltage, increasing the switches V_{GS} voltage. This increase will decrease the switches
resistance and its variation. In order to obtain an approximate value for the output
voltage of the clock boost circuit, the charge equation of the circuit must be calculated
considering all the parasitic capacitances (Fig. 5.6).

Replacing the capacitors that are connected in parallel with an equivalent
capacitor, the circuit can be simplified into the one shown in Fig. 5.7, where:

$$\begin{cases} C_{eq1} = C_{Ps4} + C_{Ps2} \\ C_{eq2} = C_{M1} + C_{Ns5a} + C_{Nd5a} + C_{Ns5} + C_{Pd2} + C_{Pg4} + C_{Ng6} + C_{Ng1} \\ C_{eq3} = C_{Nd6s} + C_{Nd6} + C_{Pd6} + C_{Pg2} \\ C_{eq4} = C_{Nd3} + C_{Ns6s} + C_{Ns6} + C_{Ns1} \end{cases} \qquad (5.5)$$

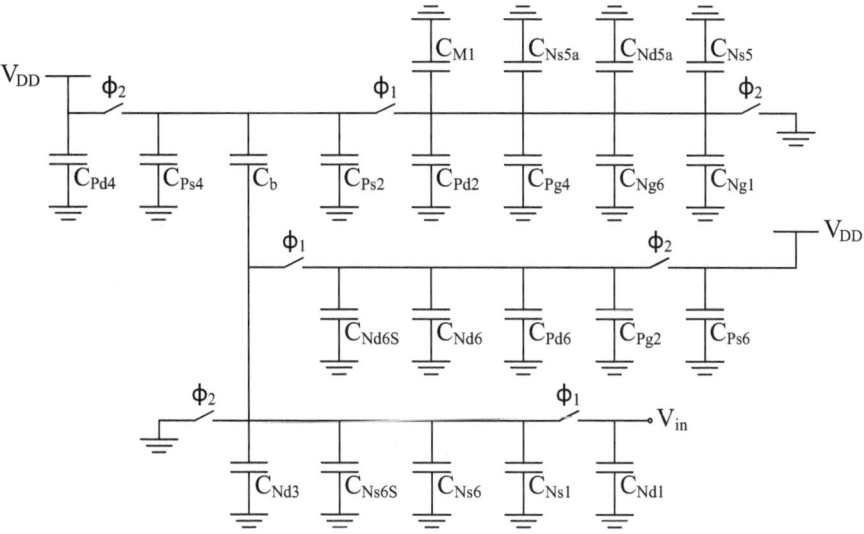

Fig. 5.6 Clock boost circuit considering parasitic capacitances

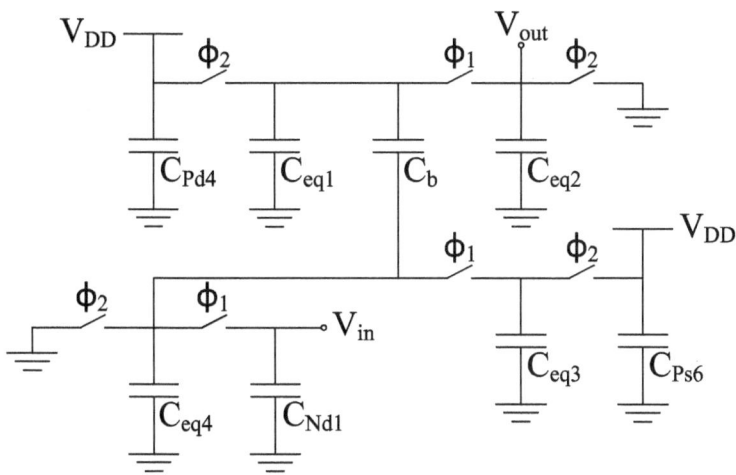

Fig. 5.7 Simplified clock boost circuit considering parasitic capacitances

Figure 5.8a and b show the equivalent circuit during clock phase ϕ_1 and clock phase ϕ_2, respectively.

Assuming charge conservation between phases, the output voltage considering parasitics (Eq. 5.6) is obtained.

$$V_{out} = \frac{V_{DD}(C_b + C_{eq1}) + V_{in}C_b}{C_b + C_{eq1} + C_{eq2}} \qquad (5.6)$$

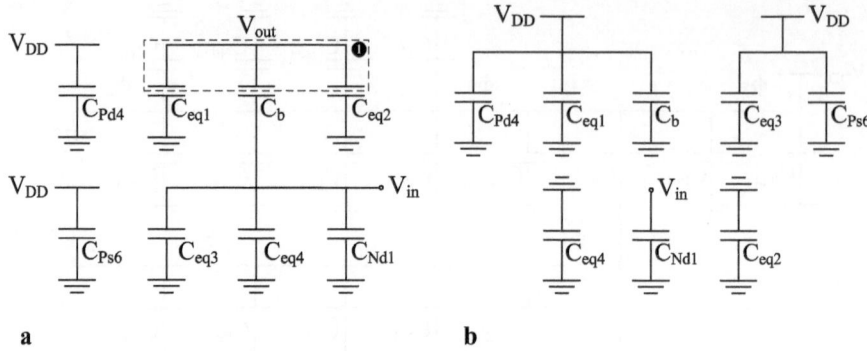

Fig. 5.8 Simplified clock boost circuit during: **a** Clock phase 1 **b** Clock phase 2

Table 5.3 Simulated clock boost circuit parasitic capacitance values

Parasitic	Transistors								
Capacitances	N1	P2	N3	P4	N5	N5a	N6	N6S	P6
C_g (aF)	660.8	3073.5	705.6	3237.6	705.6	656.7	660.8	705.7	5709.3
C_d (aF)	290.0	1240.0	290.0	1578.2	290.1	292.2	290.0	290.0	7173.9
C_s (aF)	290.0	1240.4	290.1	1651.7	290.0	292.4	290.0	290.0	5221.5

5.2.2 Simulation Results

The clock boost circuit was simulated using NMOS transistors with minimum width
and length (W = 160 nm and L = 120 nm) and PMOS transistors with four times
the NMOS width (W = 640 nm and L = 120 nm). The capacitor C_b was made using
a PMOS transistor (L = W = 4 μm) in which the source, drain and bulk nodes are
connected together. The equivalent capacitance value is approximately 145 fF and
was chosen so that it is considerably larger than the parasitic capacitances in order to
minimize their influence. Alternatively a MIMCAP capacitor could have been used
as the capacitor C_b, although to achieve the same capacitance value the capacitor
would have to be three times the size of the PMOS transistor used.

Figure 5.9 shows clock phase 1 and the boosted phase using the clock boost
circuit. The output voltage was boosted from 1.2 to 2.216 V. Using Eq. 5.6 with the
simulated parasitic capacitance values (Table 5.3) the expected output voltage would
be 2.274 V.

Figure 5.10 shows the conductance of a switch in function of the input voltage
with and without the use of a clock boost circuit. It can be concluded that when using
the clock boost circuit the conductance increases and its non-linearity decreases
when compared to not using a clock boost circuit. The decrease in non-linearity will
substantially decrease the distortion introduced by the switches, as it was seen in the
previous section (Table 5.2).

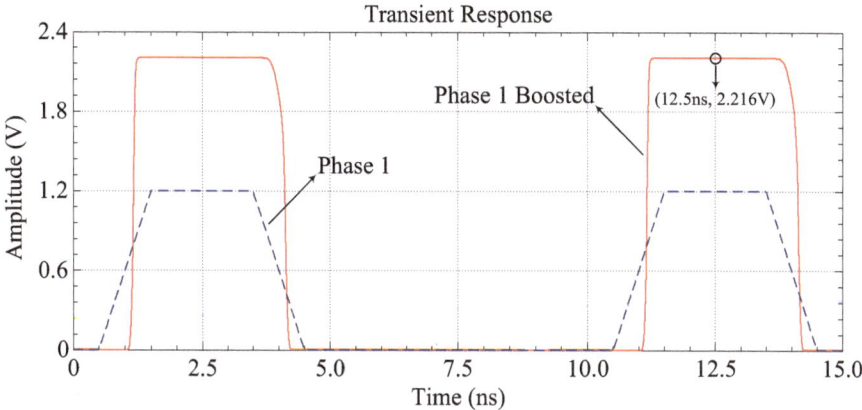

Fig. 5.9 Clock boosted phase 1

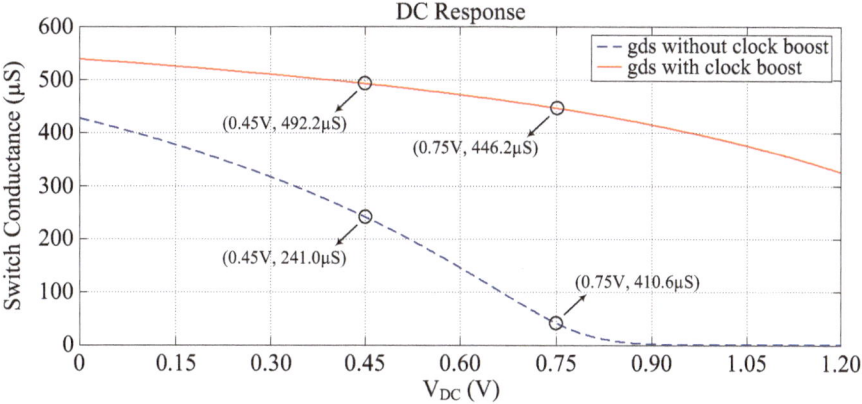

Fig. 5.10 Effect of clock boosting on the switch's conductance

5.3   Low-Voltage Clock Boost Circuit

In this section an alternative clock boost circuit is analyzed and simulated. This circuit is capable of generating a voltage level higher than the chips supply voltage and can operate with supply voltages lower than 1.2 V.

5.3.1   Low-Voltage Clock Boost Circuit Analysis

The practical realization of the clock boost circuit is shown in Fig. 5.11. This circuit, using clock phases ϕ_1 and ϕ_2, generates boosted phases ϕ_{1_B} and ϕ_{2_B} that can be used to drive NMOS switches, even at low-voltage.

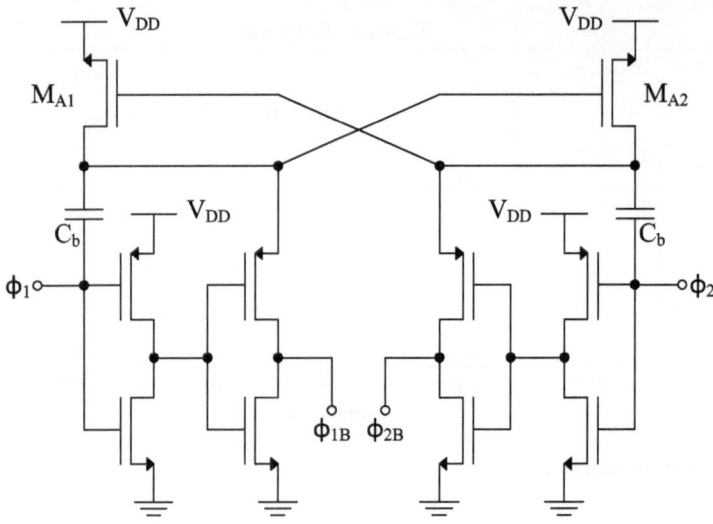

Fig. 5.11 Clock boost circuit [11]

As it was described in Sect. 5.2.1 the output voltage of the clock boost circuit is dependent on the circuits parasitic capacitances. In order to reduce the influence of these capacitances on the circuits output voltage, capacitor C_b has to be considerably larger than the parasitic capacitances.

5.3.2   Simulation Results

The clock boost circuit was simulated using NMOS transistors with minimum width and length (W = 160 nm and L = 120 nm) and PMOS transistors with three times the NMOS width (W = 480 nm and L = 120 nm). The capacitor C_b was implemented with a MIMCAP capacitor (L = 34.38 μm and W = 20 μm) with an equivalent capacitance value of 700 fF in order to minimize the influence of the parasitic capacitances.

In [11] the clock boost circuit was implemented with a supply voltage of 250 mV and was able to operate up to a clock speed of 1.4 MHz. It was also concluded in this paper that increasing the supply voltage to 300 mV would allow the clock speed to be increased to 2.8 MHz. Figure 5.12 shows one of the problems of increasing the clock speed without increasing the supply voltage. With the increase in clock speed the deformation in the falling edge of the clock signal increases making the clock signal take longer to return to the 'low' value, starting to overlap.

The clock boost circuit was simulated for different supply voltages, between 1.2 and 0.5 V, for a clock speed of 10 MHz in order to determine the voltage at which this circuit offers the best performance (Fig. 5.13). In this case, lowering the supply voltage increases the voltage boosting factor of the circuit.

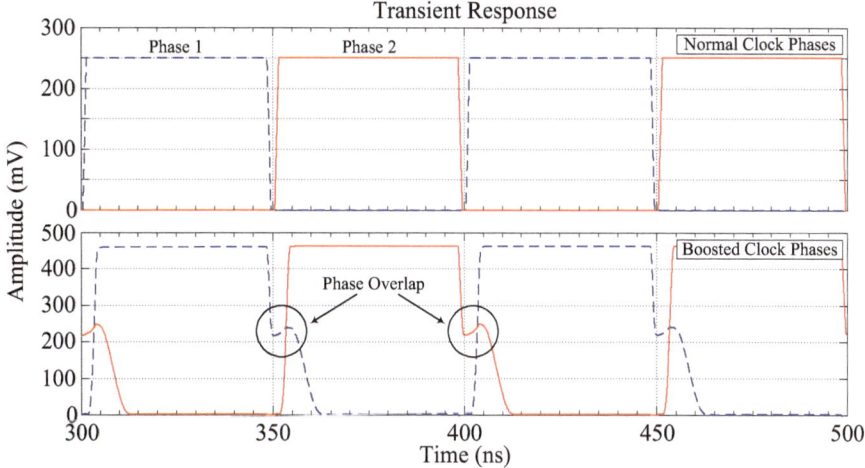

Fig. 5.12 Phase overlap due to clock speed increase beyond the circuit bandwidth

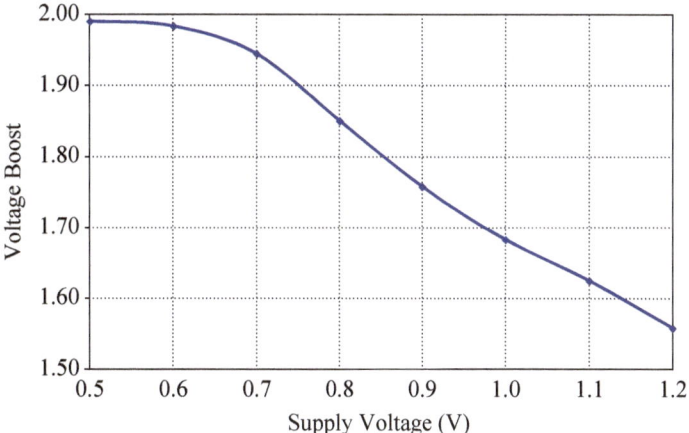

Fig. 5.13 Voltage boosting at 10 MHz

Figure 5.14 shows clock phase 1 and the boosted phase using the clock boost circuit for three different supply voltages raging from 1.2 to 0.5 V. With the decrease in supply voltage, the voltage boost increases and the clock signal becomes closer to an ideal clock signal.

Figure 5.15 shows the conductance of a switch in function of the input voltage with and without the use of a clock boost circuit for three different supply voltages. It can be concluded that with the decrease in supply voltage the conductance also decreases, which means that the switches resistance increases and it is expected that the distortion introduced by the switches increase.

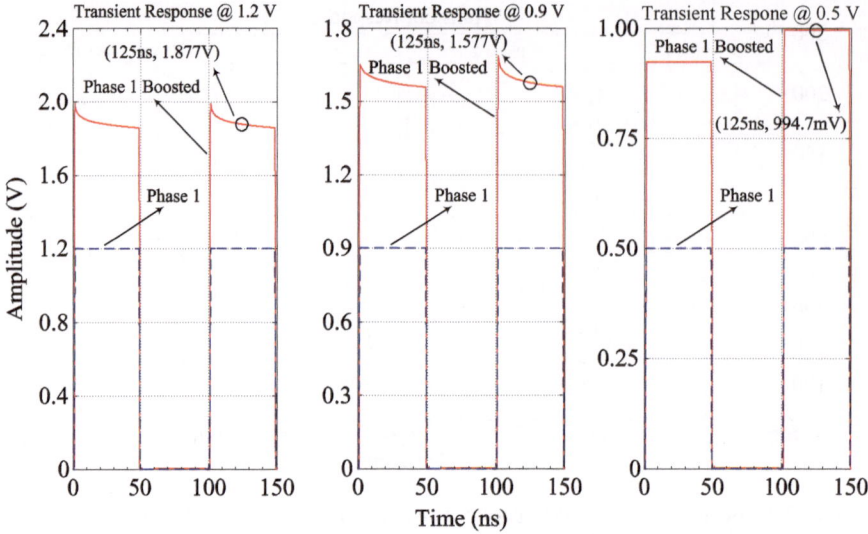

Fig. 5.14 Clock boosted phase 1 for three different supply voltages

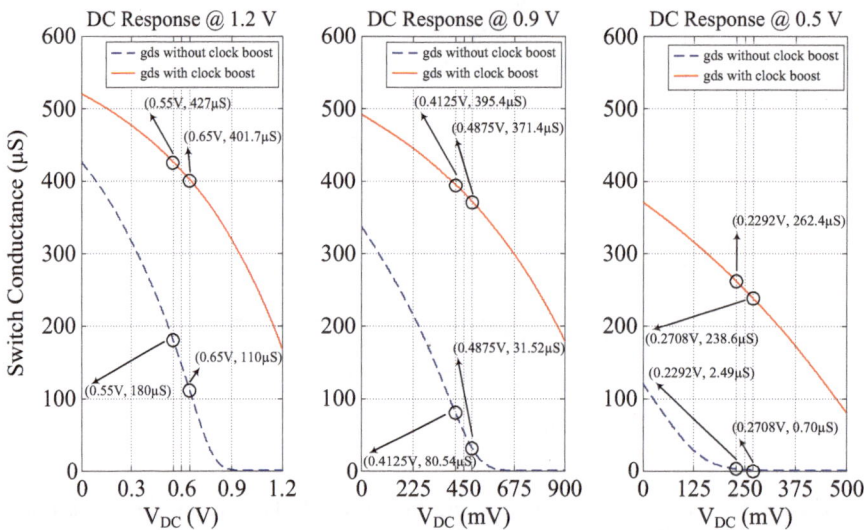

Fig. 5.15 Effect of clock boosting on the switches conductance for three different supply voltages

The distortion results are shown in Fig. 5.16 and were obtained from simulating the clock boost circuit in the second-order low-pass SC filter (Fig. 3.6) with ideal components except for the switches. All switches were simulated using minimum length (L = 120 nm).

Besides the conclusions that were obtained in Sect. 5.2.2 concerning the distortion, it can also be concluded with the simulation results obtained here that, although the

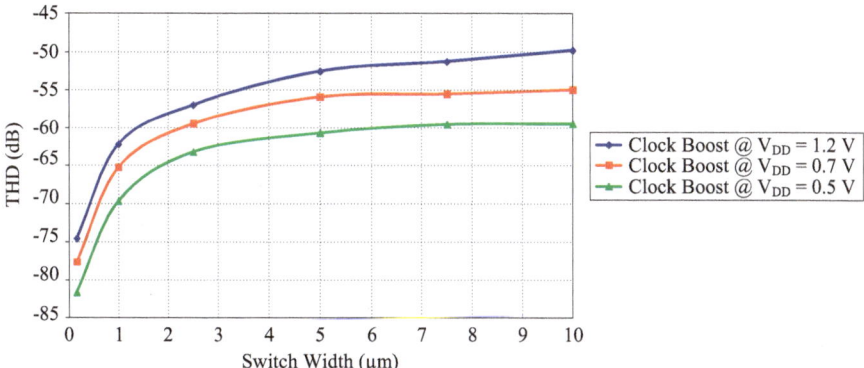

Fig. 5.16 Distortion with and without clock boosting in function of the switches width

switches resistance increases with the decrease in supply voltage, the distortion in the filter decreases. This decrease in distortion is due to the non-linear effects of the parasitic capacitances, which decrease with the supply voltage, while the non-linear effects of the switches resistance remains relatively constant.

5.4   Source Follower with g_{ds} Compensation

The source follower with g_{ds} compensation, which has the basic topology shown in Fig. 5.17, will be used in the implementation of the low-pass filters. In this section three versions of this buffer will be analyzed in order to determine which one offers the best performance: with body effect only in transistor M_2 (henceforth called source follower with g_{ds} compensation), with body effect only in transistor M_1 (complementary source follower with g_{ds} compensation), and with both transistors M_1 and M_2 without body effect (source follower with g_{ds} and body effect compensation).

Considering a follower circuit formed only by transistors M_1 and M_3, and that M_3 is an ideal current source with infinite impedance, the gain of the circuit is $G = g_{m_1}/(g_{m_1} + g_{ds_1})$. Since in modern nanometer technologies the ratio g_m/g_{ds} is below 10, the gain of the follower is below 0.91. In order to improve the gain of the buffer, a second follower is used (transistor M_2 and M_4) to compensate the effect of g_{ds_1}. This follower will make it so that the voltage in the drain of transistor M_1 will vary depending on the input voltage V_{in}. Since the voltage in the source of transistor M_1 also varies depending on the input voltage V_{in}, less current will flow throw r_{ds_1} and the gain of the circuit will increase.

5.4.1   Source Follower with g_{ds} Compensation

This version of the source follower with g_{ds} compensation has body effect only in transistor M_2. The representation of this circuit is shown in Fig. 5.18.

Fig. 5.17 Basic buffer
topology

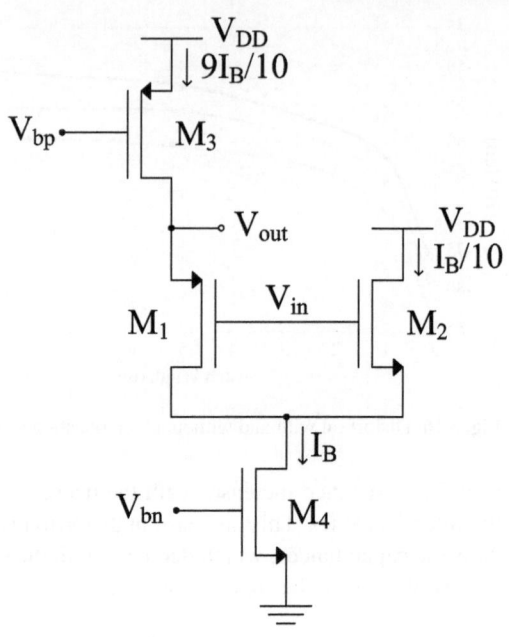

Fig. 5.18 Source follower
with g_{ds} compensation

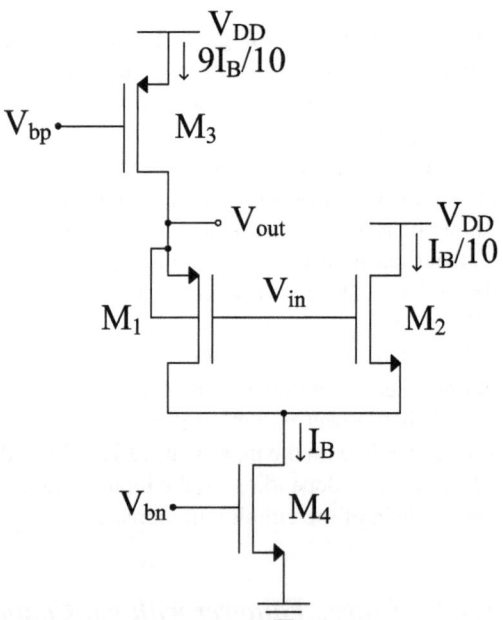

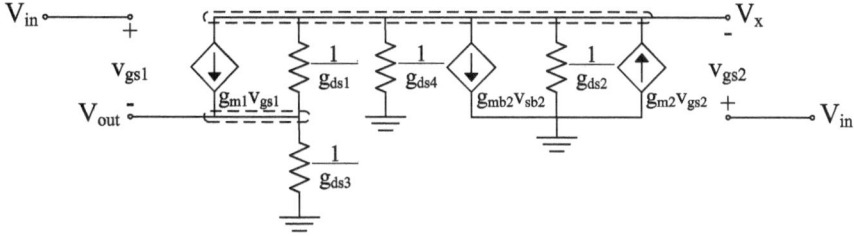

Fig. 5.19 Source follower with g_{ds} compensation low frequency small signal model

5.4.1.1 Low Frequency Small Signal Model

The low frequency small signal model of this buffer is shown in Fig. 5.19. Since this buffer will be implemented using N(P)_12_HSL130E transistor models, this means that transistor M_2 will have body effect since in standard CMOS technology the bulk node of NMOS transistors cannot be accessed.

Knowing that $V_{gs_1} = V_{in} - V_{out}$, $V_{gs_2} = V_{in} - V_x$, and $V_{sb_2} = V_x$, and applying KCL in node V_{out} and node V_x, the equations in Eq. 5.7 are obtained.

$$\begin{cases} V_{out}(g_{m_1} + g_{ds_1} + g_{ds_3}) = V_{in}g_{m_1} + V_x g_{ds_1} \\ V_x(g_{m_2} + g_{mb_2} + g_{ds_1} + g_{ds_2} + g_{ds_4}) = V_{in}(-g_{m_1} + g_{m_2}) + V_{out}(g_{m_1} + g_{ds_1}) \end{cases}$$

$$(5.7)$$

Combining both equations (Eq. 5.7) the gain equation is obtained (Eq. 5.8).

$$G = \frac{g_{m_2}g_{ds_1} + g_{m_1}(g_{m_2} + g_{mb_2} + g_{ds_2} + g_{ds_4})}{(g_{m_1} + g_{ds_3})(g_{m_2} + g_{mb_2} + g_{ds_2} + g_{ds_4}) + g_{ds_1}(g_{m_2} + g_{mb_2} + g_{ds_2} + g_{ds_3} + g_{ds_4})}$$

$$(5.8)$$

5.4.1.2 Transistor Thermal Noise Referred to the Input

The thermal noise in transistors can be represented by a current source between the drain and the source. The noise referred to the input can be represented by Eq. 5.9,

$$\overline{V_{n_{in}}^2} = \frac{\overline{V_n^2}}{G^2}$$

$$(5.9)$$

where for the buffer in question,

$$\overline{V_n^2} = (K_1 I_{n_1})^2 + (K_2 I_{n_2})^2 + (K_3 I_{n_3})^2 + (K_4 I_{n_4})^2$$

$$(5.10)$$

The coefficients K_1, K_2, K_3, and K_4 are equations in term of g_m and g_{ds}, and $I_{n_x} = 4\gamma K T g_{m_x}$. In order to obtain these coefficients the gain equation must be recalculated from the small signal model considering the thermal noise sources (Fig. 5.20).

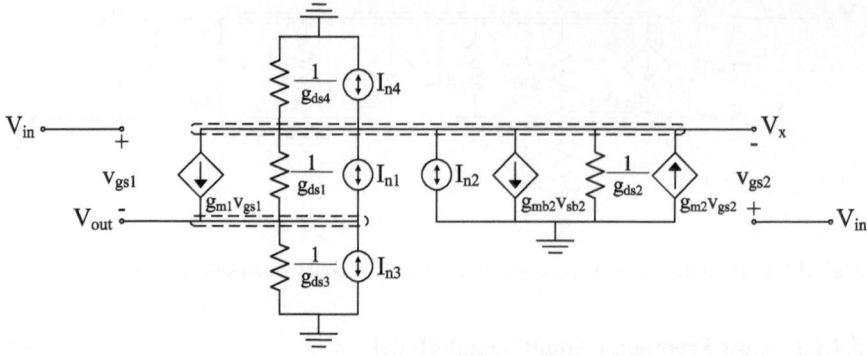

Fig. 5.20 Source follower with g_{ds} compensation low frequency small signal model considering thermal noise

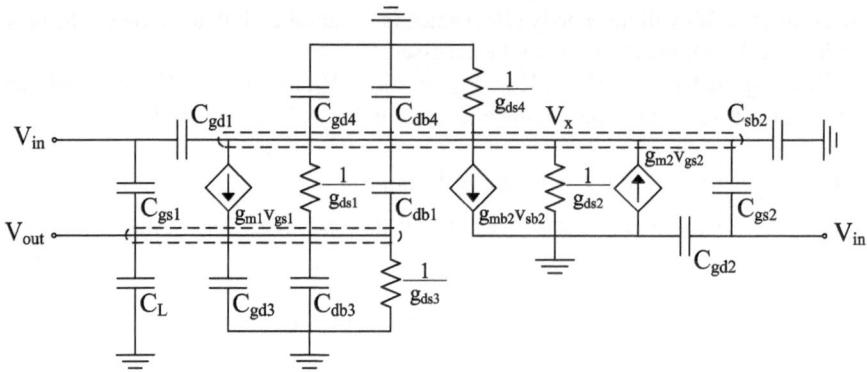

Fig. 5.21 Source follower with g_{ds} compensation medium frequency small signal model

The equations in Eq. 5.11 are obtained from the small signal model shown in Fig. 5.20. In order to obtain the V_n equation, the equations must be combined and V_{in} should be equal to 0.

$$\begin{cases} V_{out}(g_{m_1} + g_{ds_1} + g_{ds_3}) = V_{in}g_{m_1} + V_x g_{ds_1} + I_{n_1} + I_{n_3} \\ V_x(g_{m_2} + g_{mb_2} + g_{ds_1} + g_{ds_2} + g_{ds_4}) = \\ = V_{in}(-g_{m_1} + g_{m_2}) + V_{out}(g_{m_1} + g_{ds_1}) + I_{n_1} + I_{n_2} + I_{n_4} \end{cases} \quad (5.11)$$

5.4.1.3   Medium Frequency Small Signal Model

The medium frequency small signal model of this buffer is shown in Fig. 5.21.

Fig. 5.22 Complementary
source follower with g_{ds}
compensation

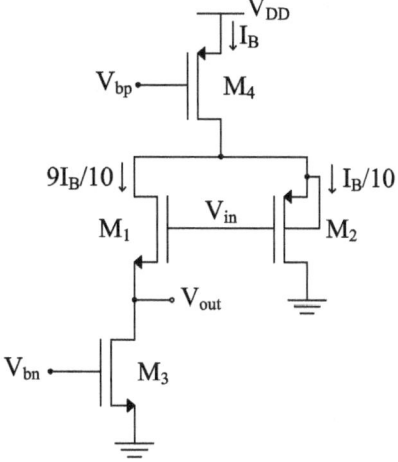

Applying KCL in node V_{out} and node V_x, the equations in Eq. 5.12 are obtained.

$$
\begin{cases}
V_{out}(g_{m_1} + g_{ds1} + g_{ds3} + s(C_L + C_{db_1} + C_{gs_1} + C_{db_3} + C_{gd_3})) = \\
\quad = V_{in}(g_{m_1} + sC_{gs_1}) + V_x(g_{ds_1} + sC_{db_1}) \\
V_x(g_{m_2} + g_{mb_2} + g_{ds_1} + g_{ds_2} + g_{ds_4} + s(C_{db_1} + C_{gd_1} + C_{gs_2} + C_{sb_2} + C_{db_4} + \\
\quad + C_{gd_4})) = V_{in}(-g_{m_1} + g_{m_2} + s(C_{gd_1} + C_{gs_2})) + V_{out}(g_{m_1} + g_{ds_1} + sC_{db_1})
\end{cases}
$$
$$(5.12)$$

Combining the equations in Eq. 5.12 a transfer function with two poles and two zeros is obtained. Due to its size the transfer function is not presented here.

5.4.2   Complementary Source Follower with g_{ds} Compensation

This version of the complementary source follower with g_{ds} compensation has body effect only in transistor M_1. The representation of this circuit is shown in Fig. 5.22.

5.4.2.1   Low Frequency Small Signal Model

The low frequency small signal model of this buffer is shown in Fig. 5.23.

Knowing that $V_{gs_1} = V_{in} - V_{out}$, $V_{gs_2} = V_{in} - V_x$, and $V_{sb_1} = V_{out}$, and applying KCL in node V_{out} and node V_x, the equations in Eq. 5.13 are obtained.

$$
\begin{cases}
V_{out}(g_{m_1} + g_{mb_1} + g_{ds_1} + g_{ds_3}) = V_{in}g_{m_1} + V_x g_{ds_1} \\
V_x(g_{m_2} + g_{ds_1} + g_{ds_2} + g_{ds_4}) = V_{in}(-g_{m_1} + g_{m_2}) + V_{out}(g_{m_1} + \\
\quad + g_{mb_1} + g_{ds_1})
\end{cases}
\qquad (5.13)
$$

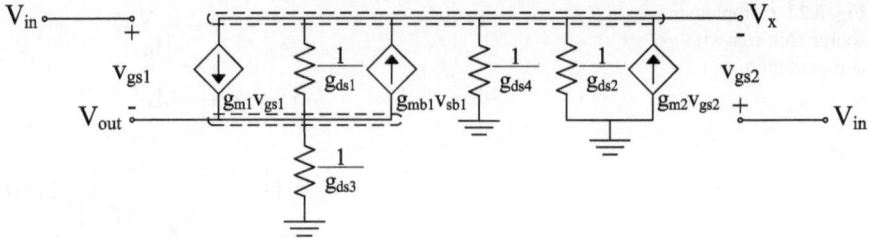

Fig. 5.23 Complementary source follower with g_{ds} compensation low frequency small signal model

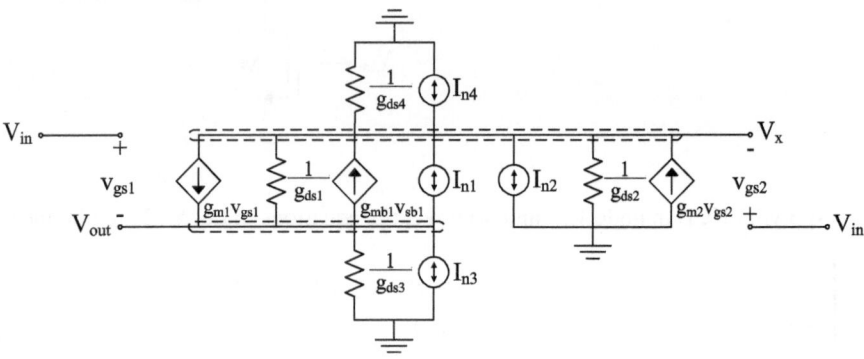

Fig. 5.24 Complementary source follower with g_{ds} compensation low frequency small signal model considering thermal noise

Combining both equations (Eq. 5.13) the gain equation is obtained (Eq. 5.14).

$$G = \frac{g_{m2}g_{ds_1} + g_{m_1}(g_{m_2} + g_{ds_2} + g_{ds_4})}{g_{ds_1}(g_{m_2} + g_{ds_2} + g_{ds_3} + g_{ds_4}) + (g_{m_2} + g_{ds_2} + g_{ds_4})(g_{m_1} + g_{mb_1} + g_{ds_3})}$$
$$(5.14)$$

5.4.2.2 Transistor Thermal Noise Referred to the Input

The thermal noise referred to the input can be calculated using the process described in Sect. 5.4.1.2. The small signal model considering the thermal noise sources is shown in Fig. 5.24.

The equations in Eq. 5.15 are obtained from the small signal model shown in Fig. 5.24. In order to obtain the V_n equation, the equations must be combined and V_{in} should be considered equal to 0.

$$\begin{cases} V_{out}(g_{m_1} + g_{mb_1} + g_{ds_1} + g_{ds_3}) = V_{in}g_{m_1} + V_x g_{ds_1} + I_{n_1} + I_{n_3} \\ V_x(g_{m_2} + g_{ds_1} + g_{ds_2} + g_{ds_4}) = V_{in}(-g_{m_1} + g_{m_2}) + V_{out}(g_{m_1} + \\ + g_{mb_1} + g_{ds_1}) + I_{n_1} + I_{n_2} + I_{n_4} \end{cases} \quad (5.15)$$

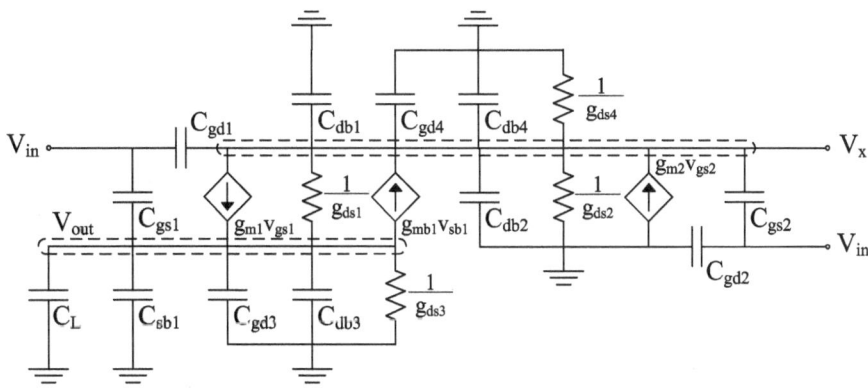

Fig. 5.25 Complementary source follower with g_{ds} compensation medium frequency small signal model

5.4.2.3 Medium Frequency Small Signal Model

The medium frequency small signal model of this buffer is shown in Fig. 5.25.

Applying KCL in node V_{out} and node V_x, the equations in Eq. 5.16 are obtained.

$$\begin{cases} V_{out}(g_{m_1} + g_{mb_1} + g_{ds_1} + g_{ds_3} + s(C_L + C_{gs_1} + C_{sb_1} + C_{db_3} + C_{gd_3})) = \\ = V_{in}(g_{m_1} + sC_{gs_1}) + V_x g_{ds_1} \\ V_x(g_{m_2} + g_{ds_1} + g_{ds_2} + g_{ds_4} + s(C_{db_1} + C_{gd_1} + C_{db_2} + C_{gs_2} + C_{db_4} + \\ + C_{gd_4})) = V_{in}(-g_{m_1} + g_{m_2} + s(C_{gd_1} + C_{gs_2})) + V_{out}(g_{m_1} + g_{mb_1} + g_{ds_1}) \end{cases}$$
$$(5.16)$$

Combining the equations in Eq. 5.16 a transfer function with two poles and two zeros is obtained. Due to its size the transfer function is not presented here.

5.4.3 Source Follower with g_{ds} and Body Effect Compensation

The source follower with g_{ds} and body effect compensation has the bulks of transistor M_1 and M_2 connected to their respective sources, eliminating the body effect. Both representations of this circuit are shown in Fig. 5.26. To eliminate the body effect in both transistors, the buffer was implemented using N(P)_12_RF transistor models. The equations presented in this section are valid for both versions of the buffer.

5.4.3.1 Low Frequency Small Signal Model

The low frequency small signal model of this buffer is shown in Fig. 5.27.

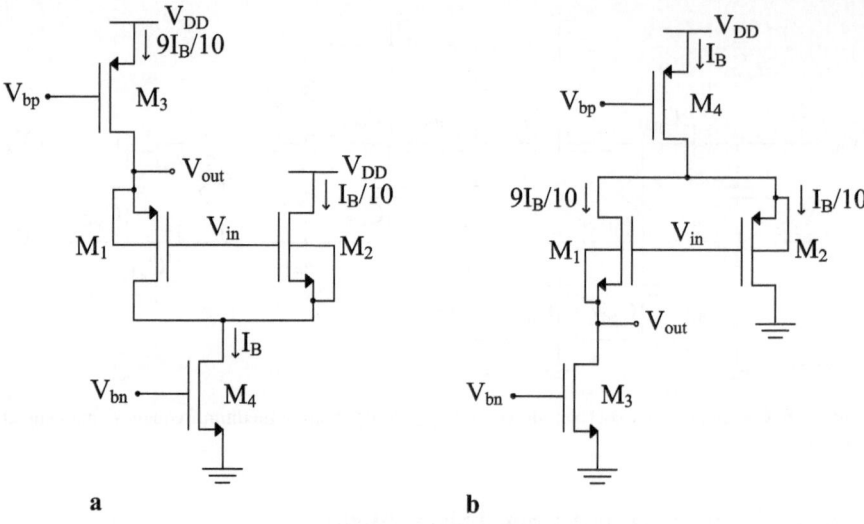

Fig. 5.26 Source follower with g_{ds} and body effect compensation: **a** Normal version **b** Complementary version

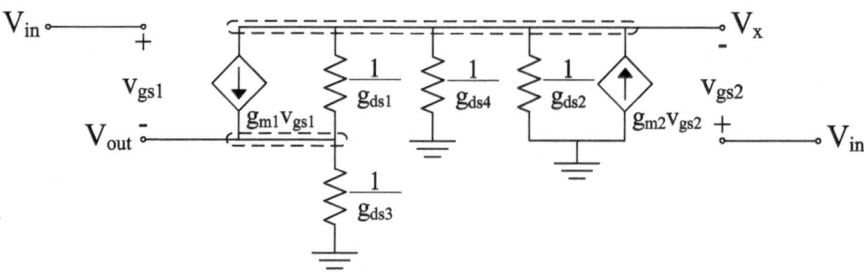

Fig. 5.27 Source follower with g_{ds} and body effect compensation low frequency small signal model

Knowing that $V_{gs_1} = V_{in} - V_{out}$, and $V_{gs_2} = V_{in} - V_x$, and applying KCL in node V_{out} and node V_x, the equations in Eq. 5.17 are obtained.

$$\begin{cases} V_{out}(g_{m_1} + g_{ds_1} + g_{ds_3}) = V_{in}g_{m_1} + V_x g_{ds_1} \\ V_x(g_{m_2} + g_{ds_1} + g_{ds_2} + g_{ds_4}) = V_{in}(-g_{m_1} + g_{m_2}) + V_{out}(g_{m_1} + g_{ds_1}) \end{cases} \quad (5.17)$$

Combining both equations (Eq. 5.17) the gain equation is obtained (Eq. 5.18).

$$G = \frac{g_{m_2}g_{ds_1} + g_{m_1}(g_{m_2} + g_{ds_2} + g_{ds_4})}{(g_{m_1} + g_{ds_3})(g_{m_2} + g_{ds_2} + g_{ds_4}) + g_{ds_1}(g_{m_2} + g_{ds_2} + g_{ds_3} + g_{ds_4})} \quad (5.18)$$

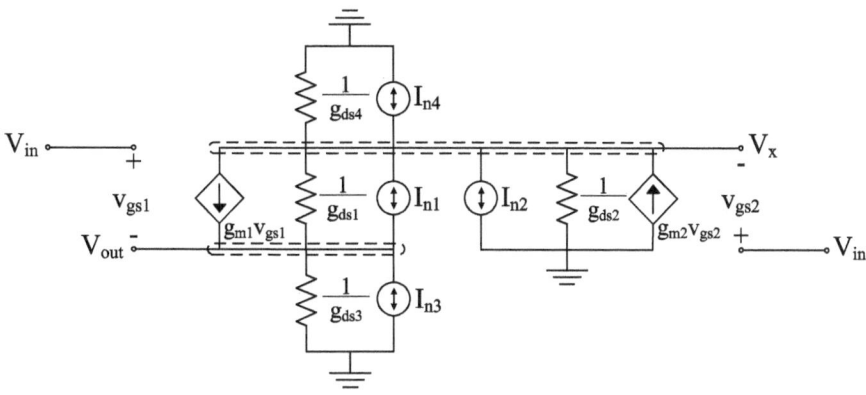

Fig. 5.28 Source follower with g_{ds} and body effect compensation low frequency small signal model considering thermal noise

5.4.3.2   Transistor Thermal Noise Referred to the Input

The thermal noise referred to the input can be calculated using the process described in Sect. 5.4.1.2. The small signal model considering the thermal noise sources is shown in Fig. 5.28.

The equations in Eq. 5.19 are obtained from the small signal model shown in Fig. 5.28. In order to obtain the V_n equation, the equations must be combined and V_{in} should be considered equal to 0.

$$
\begin{cases}
V_{out}(g_{m_1} + g_{ds_1} + g_{ds_3}) = V_{in}g_{m_1} + V_x g_{ds_1} + I_{n_1} + I_{n_3} \\
V_x(g_{m_2} + g_{ds_1} + g_{ds_2} + g_{ds_4}) = V_{in}(-g_{m_1} + g_{m_2}) + V_{out}(g_{m_1} + g_{ds_1}) + \\
+ I_{n_1} + I_{n_2} + I_{n_4}
\end{cases}
$$

$$(5.19)$$

5.4.3.3   Medium Frequency Small Signal Model

The medium frequency small signal model of this buffer is shown in Fig. 5.29.

Applying KCL in node V_{out} and node V_x, the equations in Eq. 5.20 are obtained.

$$
\begin{cases}
V_{out}(g_{m_1} + g_{ds_1} + g_{ds_3} + s(C_L + C_{gs_1} + C_{db_1} + C_{db_3} + C_{gd_3})) = \\
= V_{in}(g_{m_1} + sC_{gs_1}) + V_x(g_{ds_1} + sC_{db_1}) \\
V_x(g_{m_2} + g_{ds_1} + g_{ds_2} + g_{ds_4} + s(C_{db_1} + C_{gd_1} + C_{db_2} + C_{gs_2} + C_{db_4} + C_{gd_4})) = \\
= V_{in}(-g_{m_1} + g_{m_2} + s(C_{gd_1} + C_{gs_2})) + V_{out}(g_{m_1} + g_{ds_1}) + sC_{db_1}
\end{cases}
$$

$$(5.20)$$

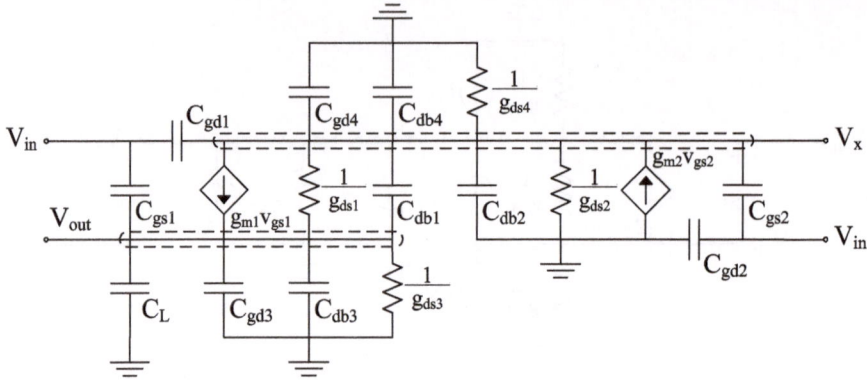

Fig. 5.29 Source follower with g_{ds} and body effect compensation medium frequency small signal model

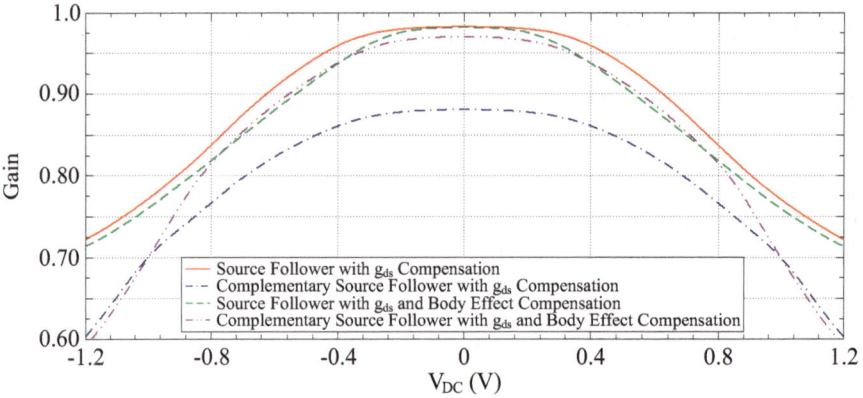

Fig. 5.30 Relationship between the gain of the circuits and the DC voltage

Combining the equations in Eq. 5.20 a transfer function with two poles and two zeros is obtained. Due to its size the transfer function is not presented here.

5.4.4 Simulation Results

The three buffers presented in the previous sections were simulated in order to determine which one has the highest gain, since there is an inverse relationship between the gain of the buffer and the size of the capacitors in the SC filter. The biasing voltages of the buffers were generated using an ideal current source connected to a transistor in diode configuration. Figure 5.30 shows the gain of the buffers as a function of the DC voltage. To obtain this relation a DC simulation was done sweeping the DC value of an ideal voltage source between 0 and 1.2 V (V_{SS} and V_{DD}). From the figure it can be concluded that the buffers have better linearity at half the supply voltage (0.6 V).

The current of the buffers was chosen in order to achieve a certain GBW. The value for the GBW was obtained from the equation in Eq. 5.21 using the filters clock frequency [12].

$$e^{-GBW[rad/s]T_s/2} < 0.1\% \Leftrightarrow GBW[Hz] > -\frac{\ln(0.1\%)F_s}{\pi} \qquad (5.21)$$

Assuming that the clock frequency of the filter is 100 MHz, this means that the current should be chosen so that the GBW of the buffers is higher than 220 MHz. For the normal version the current used in the simulations was 550 μA and for the complementary version 400 μA. The transistor sizes used for the simulations in this section are shown in Tables 5.4 and 5.5.

The simulation results with a common mode voltage of 0.6 V, supply voltage of 1.2 V, single-ended input signal with an amplitude of 150 mV, and a load capacitance of 4 pF are shown in Table 5.6. The figure of merit (FOM) values shown in Tables 5.6 and 5.7 were calculated using Eq. 5.22 [13].

$$FOM\left[\frac{MHz\,pF}{mW}\right] = \frac{GBW\,C_L}{Power} \qquad (5.22)$$

The results show that the THD of the buffers is relatively high (between -44 and -51 dB). One of the reasons of the high distortion is the fact that although the buffers input signal is centered at half the supply voltage, the output of the buffer has a common mode voltage value of $\frac{V_{DD}}{2} - V_{GS_1}$. Since in this case the buffers have a V_{GS_1} of around $\mp 300 mV$ and, on top of that, the input signal is sinusoidal, the signals maximum/minimum voltage, depending on the version of the buffer, reaches close to V_{DD}/V_{SS} degrading the quality of the signal. In order to improve the distortion a new common mode voltage value was calculated for the buffer according to the equation in Eq. 5.23.

$$V_{cm} = \frac{V_{DD} - V_{GS_1}}{2} \qquad (5.23)$$

The results obtained from simulating the buffers with the new common mode voltage are shown in Table 5.7. The results show that by adjusting the circuits common mode voltage it is possible to improve the distortion values by over 10 dB.

The results from Tables 5.6 and 5.7 show that the buffers (version (d)), using the RF transistor models, provide better results. For the source follower with g_{ds} compensation and the source follower with g_{ds} and body effect compensation, the gain remains relatively identical, and the GBW of the buffer sightly decreased. For the complementary source follower with g_{ds} compensation and complementary source follower with g_{ds} and body effect compensation, the GBW sightly decreases but the gain increases significantly making it easier to obtain smaller capacitance values for the SC filter.

Table 5.4 HSL130E transistor parameters: (a) Source follower (SF) with g_{ds} compensation with DC @ 0.6 V (b) Complementary source follower (CSF) with g_{ds} compensation with DC @ 0.6 V (c) SF with g_{ds} compensation with DC @ 0.45 V (d) CSF with g_{ds} compensation with DC @ 0.75 V

		$W(\mu m)$	$L(\mu m)$	$I_D(\mu A)$	$g_m(mS)$	$g_{ds}(\mu S)$	$V_{DSAT}(mV)$
(a)	M_1	499.20	0.24	-570.04	10.72	238.48	-86.56
	M_2	7.86	0.12	67.42	1.29	92.68	62.14
	M_3	187.20	0.36	-570.04	6.44	132.33	-153.42
	M_{3B}	18.72	0.36	-58.50	0.66	9.48	-153.84
	M_4	374.40	0.36	637.46	14.02	289.33	67.70
	M_{4B}	37.44	0.36	65.00	1.43	27.54	67.70
(b)	M_1	130.62	0.36	360.17	7.40	121.49	75.69
	M_2	28.80	0.36	-33.87	0.57	5.79	-99.64
	M_3	23.00	0.36	360.17	3.97	149.24	148.54
	M_{3B}	2.30	0.36	36.00	0.40	10.18	146.47
	M_4	288.00	0.36	-394.04	6.12	89.56	-109.13
	M_{4B}	28.80	0.36	-40.00	0.62	7.45	-109.24
(c)	M_1	499.20	0.24	-585.61	10.92	244.71	-87.71
	M_2	7.86	0.12	23.60	0.51	35.16	50.23
	M_3	187.20	0.36	-585.61	6.62	94.36	-153.86
	M_{3B}	18.72	0.36	-58.50	0.66	9.48	-153.84
	M_4	374.40	0.36	609.20	13.39	402.80	67.70
	M_{4B}	37.44	0.36	65.00	1.43	27.54	67.70
(d)	M_1	130.62	0.36	376.53	7.67	130.03	77.78
	M_2	28.80	0.36	-8.37	0.19	1.64	-62.49
	M_3	23.00	0.36	376.53	4.15	109.30	148.55
	M_{3B}	2.30	0.36	36.00	0.40	10.18	146.47
	M_4	288.00	0.36	-384.90	5.95	169.85	-108.93
	M_{4B}	28.80	0.36	-40.00	0.62	7.45	-109.24

5.5 Low-Voltage Fully-Differential Voltage-Combiner

The fully-differential voltage-combiner amplifier, which has the basic single-ended topology shown in Fig. 5.31, will be used in the implementation of the band-pass filters and of the low-voltage low-pass filter. In this section two versions of this amplifier will be analyzed: with body effect only in transistor M_1 (henceforth called voltage-combiner), and with body effect only in transistor M_2 (complementary voltage-combiner).

In single-ended configuration this amplifier acts as a buffer and has a gain below unity. Doubling the circuit and connecting the sources of transistors M_1 (node V_{dif}), a differential pair is formed and the circuit becomes a low gain amplifier.

Table 5.5 RF_12 transistor parameters: (a) Source follower with g_{ds} compensation with DC @ 0.6 V (b) Complementary source follower with g_{ds} compensation with DC @ 0.6 V (c) Source follower with g_{ds} compensation with DC @ 0.45 V (d) Complementary source follower with g_{ds} compensation with DC @ 0.75 V

		$W(\mu m)$	$L(\mu m)$	Fingers	Multiplier	$I_D(\mu A)$	$g_m(mS)$	$g_{ds}(\mu S)$	$V_{DSAT}(mV)$
(a)	M_1	7.80	0.24	32	2	−601.69	9.10	275.57	−86.38
	M_2	1.39	0.12	4	1	78.85	1.25	79.61	94.50
	M_3	5.85	0.36	32	1	−601.69	6.68	92.12	−175.63
	M_{3B}	2.34	0.36	8	1	−58.50	0.65	6.61	−172.13
	M_4	5.85	0.36	16	4	680.54	14.62	325.23	75.32
	M_{4B}	4.68	0.36	8	1	65.00	1.42	24.24	73.61
(b)	M_1	5.76	0.36	16	1	384.09	7.08	78.80	94.15
	M_2	1.80	0.36	16	1	−30.55	0.46	6.98	−103.46
	M_3	2.30	0.36	10	1	384.09	3.92	179.82	168.56
	M_{3B}	1.15	0.36	4	1	72.00	0.79	21.31	160.57
	M_4	9.00	0.36	32	1	−414.63	6.25	107.83	−123.36
	M_{4B}	1.80	0.36	16	1	40.00	0.59	8.79	−122.97
(c)	M_1	7.80	0.24	32	2	−612.64	9.35	289.98	−88.30
	M_2	1.39	0.12	4	1	22.57	0.45	28.26	68.10
	M_3	5.85	0.36	32	1	−612.64	6.71	68.07	−175.64
	M_{3B}	2.34	0.36	8	1	−58.50	0.65	6.61	−172.13
	M_4	5.85	0.36	16	4	635.21	13.62	548.91	74.77
	M_{4B}	4.78	0.36	8	1	65.00	1.43	24.29	73.18
(d)	M_1	5.76	0.36	16	1	405.68	7.32	96.86	96.67
	M_2	1.80	0.36	16	1	−4.73	0.09	2.03	−52.02
	M_3	2.30	0.36	10	1	405.68	4.17	127.87	168.68
	M_{3B}	1.15	0.36	4	1	72.00	0.79	21.31	160.57
	M_4	9.00	0.36	32	1	−410.41	6.23	119.01	−123.36
	M_{4B}	1.80	0.36	16	1	40.00	0.59	8.79	−122.97

5.5.1 Fully-Differential Voltage Combiner

This version of the fully-differential voltage-combiner amplifier has body effect only in transistor M_1. The single-ended representation of this circuit is shown in Fig. 5.32.

5.5.1.1 Low Frequency Small Signal Model

The low frequency small signal model of this amplifier is shown in Fig. 5.33. Since this amplifier will be implemented using N(P)_12_HSL130E transistor models, this means that transistor M_1 will have body effect.

Table 5.6 Simulation results for the buffers with a common mode voltage of 0.6 V: (a) Source follower (SF) with g_{ds} compensation (b) Complementary source follower (CSF) with g_{ds} compensation (c) SF with g_{ds} and body effect compensation (d) CSF with g_{ds} and body effect compensation

Version	Power (μW)	Gain	GBW (MHz)	THD (dB)	Thermal Noise (μV)	FOM
(a)	765	0.981	303.4	-48.17	24.2	1587
(b)	473	0.881	259.4	-51.00	28.3	2194
(c)	817	0.982	259.4	-43.90	26.5	1271
(d)	498	0.970	234.9	-47.19	42.7	1888

Table 5.7 Simulated results for the buffers with a common mode voltage of 0.45/0.75 V: (a) Source follower (SF) with g_{ds} compensation (b) Complementary source follower (CSF) with g_{ds} compensation (c) SF with g_{ds} and body effect compensation (d) CSF with g_{ds} and body effect compensation

Version	Power (μW)	Gain	GBW (MHz)	THD (dB)	Thermal Noise (μV)	FOM
(a)	731	0.979	289.0	-66.96	25.2	1581
(b)	462	0.886	255.1	-62.10	29.0	2209
(c)	762	0.974	254.3	-70.63	29.6	1334
(d)	493	0.968	222.1	-60.45	42.7	1804

Fig. 5.31 Basic single-ended topology

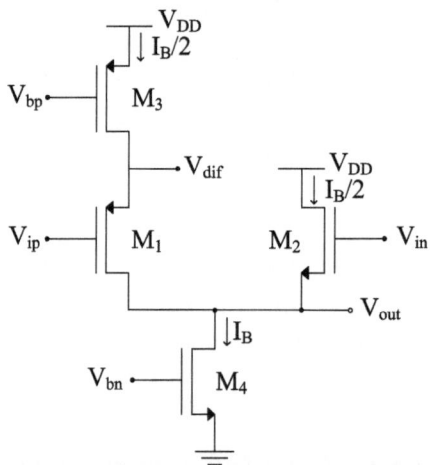

Knowing that $V_{gs_1} = V_{ip} - V_x$, $V_{gs_2} = V_{in} - V_{out}$, and $V_{sb_1} = V_x$, and applying KCL in node V_{out} and node V_x, the equations in Eq. 5.24 are obtained.

$$\begin{cases} V_{out} g_{ds_1} = -V_{ip} g_{m_1} + V_x(g_{m_1} + g_{mb_1} + g_{ds_1} + g_{ds_3}) \\ V_x(g_{m_1} + g_{mb_1} + g_{ds_1}) = V_{ip} g_{m_1} - V_{in} g_{m_2} + V_{out}(g_{m_2} + g_{ds_1} + g_{ds_2} + g_{ds_4}) \end{cases}$$

$$(5.24)$$

Fig. 5.32 Single-ended
configuration of the
voltage-combiner amplifier

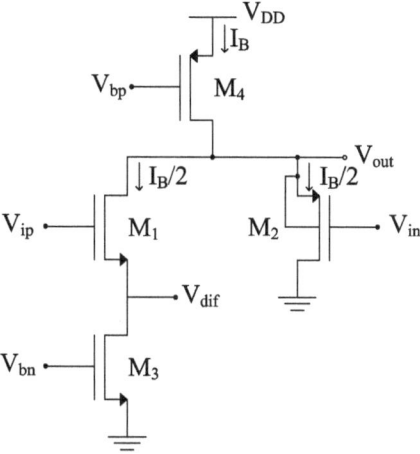

Combining both equations (Eq. 5.24) the gain equation is obtained (Eq. 5.25).

$$G = \frac{g_{m_1}g_{ds_3} + g_{m_2}(g_{m_1} + g_{mb_1} + g_{ds_1} + g_{ds_3})}{g_{ds_1}(g_{m_2} + g_{ds_2} + g_{ds_3} + g_{ds_4}) + (g_{m_2} + g_{ds_2} + g_{ds_4})(g_{m_1} + g_{mb_1} + g_{ds_3})} \tag{5.25}$$

5.5.1.2   Transistor Thermal Noise Referred to the Input

The thermal noise referred to the input can be calculated using the process described
in Sect. 5.4.1.2. The small signal model considering the thermal noise sources is
shown in Fig. 5.34.

The equations in Eq. 5.26 are obtained from the small signal model shown in
Fig. 5.34. In order to obtain the V_n equation, the equations must be combined and
V_{in} should be considered equal to 0.

$$\begin{cases} V_{out}g_{ds_1} = -V_{ip}g_{m_1} + V_x(g_{m_1} + g_{mb_1} + g_{ds_1} + g_{ds_3}) - I_{n_1} - I_{n_3} \\ V_x(g_{m_1} + g_{mb_1} + g_{ds_1}) = -V_{in}g_{m_2} + V_{out}(g_{m_2} + g_{ds_1} + g_{ds_2} + g_{ds_4}) + \\ -I_{n_1} - I_{n_2} - I_{n_4} \end{cases} \tag{5.26}$$

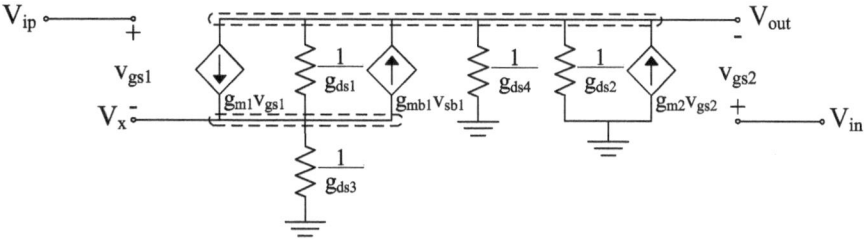

Fig. 5.33 Voltage-combiner amplifier low frequency small signal model

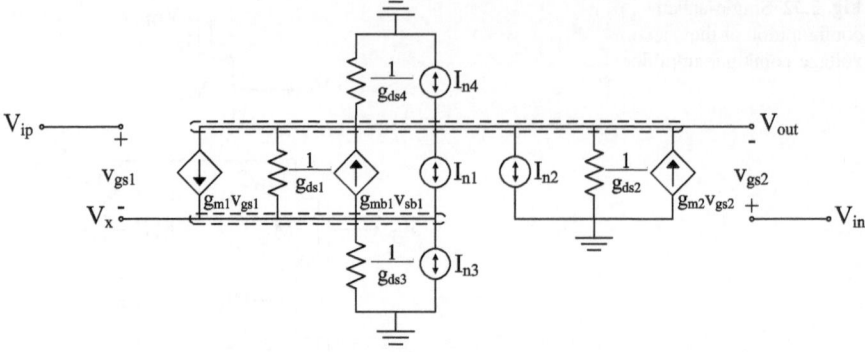

Fig. 5.34 Voltage-combiner amplifier low frequency small signal model considering thermal noise

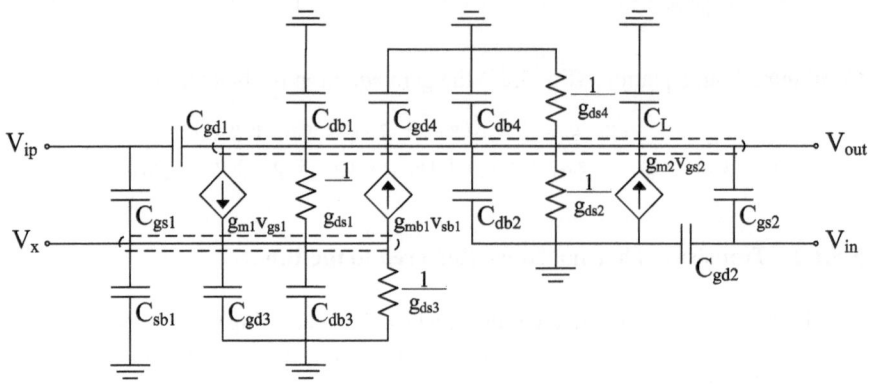

Fig. 5.35 Voltage-combiner amplifier medium frequency small signal model

5.5.1.3   Medium Frequency Small Signal Model

The medium frequency small signal model of this amplifier is shown in Fig. 5.35.
Applying KCL in node V_{out} and node V_x, the equations in Eq. 5.27 are obtained.

$$
\begin{cases}
V_{out}g_{ds_1} = -V_{ip}(g_{m_1} + sC_{gs_1}) + V_x(g_{m_1} + g_{mb_1} + g_{ds_1} + g_{ds_3} + \\
+ s(C_{gs_1} + C_{sb_1} + C_{db_3} + C_{gd_3})) \\
V_x(g_{m_1} + g_{mb_1} + g_{ds_1}) = V_{ip}(g_{m_1} - sC_{gd_1}) - V_{in}(g_{m_2} + sC_{gs_2}) + \\
+ V_{out}(g_{m_2} + g_{ds_1} + g_{ds_2} + g_{ds_4} + s(C_{db_1} + C_{gd_1} + C_{db_2} + C_{gs_2} + \\
+ C_{db_4} + C_{gd_4} + C_L))
\end{cases}
\tag{5.27}
$$

Combining the equations in Eq. 5.27 a transfer function with two poles and two
zeros is obtained. Due to its size the transfer function is not presented here.

Fig. 5.36 Single-ended
configuration of the
complementary
voltage-combiner amplifier

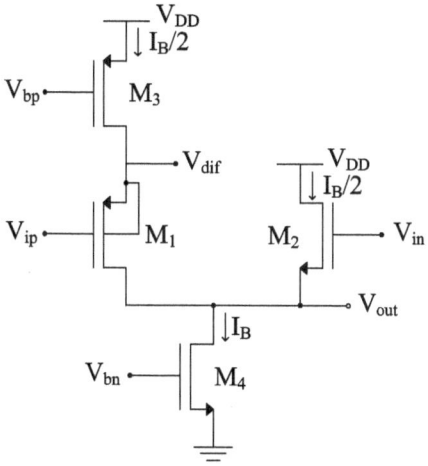

5.5.2   Complementary Fully-Differential Voltage-Combiner

The complementary voltage-combiner amplifier has body effect only in transistor
M_2. The single-ended representation of this circuit is shown in Fig. 5.36.

5.5.2.1   Low Frequency Small Signal Model

The low frequency small signal model of this amplifier is shown in Fig. 5.37.
Knowing that $V_{gs_1} = V_{ip} - V_x$, $V_{gs_2} = V_{in} - V_{out}$, and $V_{sb_2} = V_{out}$, and applying
KCL in node V_{out} and node V_x, the equations in Eq. 5.28 are obtained.

$$\begin{cases} V_{out}g_{ds_1} = -V_{ip}g_{m_1} + V_x(g_{m_1} + g_{ds_1} + g_{ds_3}) \\ V_x(g_{m_1} + g_{ds_1}) = V_{ip}g_{m_1} - V_{in}g_{m_2} + V_{out}(g_{m_2} + g_{mb_2} + g_{ds_1} + g_{ds_2} + g_{ds_4}) \end{cases}$$
$$(5.28)$$

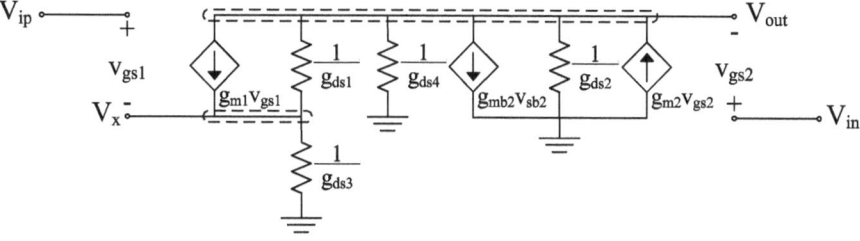

Fig. 5.37 Complementary voltage-combiner amplifier low frequency small signal model

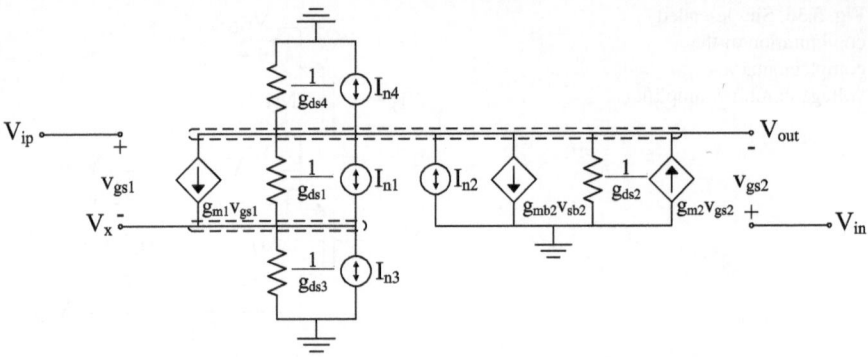

Fig. 5.38 Complementary voltage-combiner amplifier low frequency small signal model considering thermal noise

Combining both equations (Eq. 5.28) the gain equation is obtained (Eq. 5.29).

$$G = \frac{g_{m_1}g_{ds_3} + g_{m_2}(g_{m_1} + g_{ds_1} + g_{ds_3})}{(g_{m_1} + g_{ds_3})(g_{m_2} + g_{mb_2} + g_{ds_2} + g_{ds_4}) + g_{ds_1}(g_{m_2} + g_{mb_2} + g_{ds_2} + g_{ds_3} + g_{ds_4})}$$

(5.29)

5.5.2.2 Transistor Thermal Noise Referred to the Input

The thermal noise referred to the input can be calculated using the process described in Sect. 5.4.1.2. The small signal model considering the thermal noise sources is shown in Fig. 5.38.

The equations in Eq. 5.30 are obtained from the small signal model shown in Fig. 5.38. In order to obtain the V_n equation, the equations must be combined and V_{in} should be considered equal to 0.

$$\begin{cases} V_{out}g_{ds_1} = -V_{ip}g_{m_1} + V_x(g_{m_1} + g_{ds_1} + g_{ds_3}) - I_{n_1} - I_{n_3} \\ V_x(g_{m_1} + g_{ds_1}) = V_{ip}g_{m_1} - V_{in}g_{m_2} + V_{out}(g_{m_2} + g_{mb_2} + g_{ds_1} + g_{ds_2} + g_{ds_4}) + \\ -I_{n_1} - I_{n_3} - I_{n_4} \end{cases}$$

(5.30)

5.5.2.3 Medium Frequency Small Signal Model

The medium frequency small signal model of this amplifier is shown in Fig. 5.39.

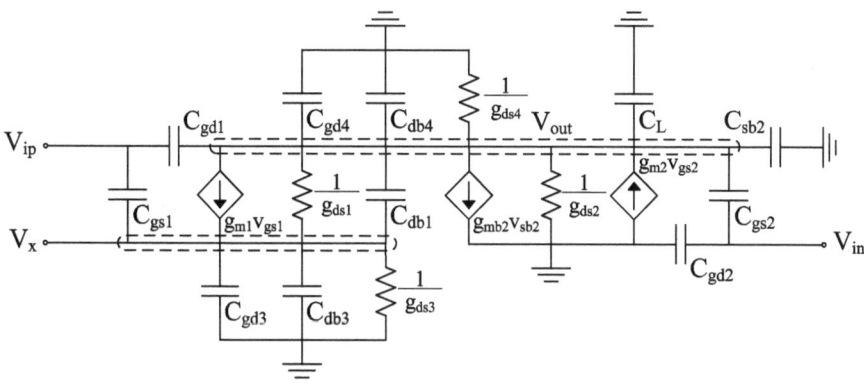

Fig. 5.39 Complementary voltage-combiner amplifier medium frequency small signal model

Applying KCL in node V_{out} and node V_x, the equations in Eq. 5.31 are obtained.

$$\begin{cases} V_{out}(g_{ds_1} + sC_{db_1}) = -V_{ip}(g_{m_1} + sC_{gs_1}) + V_x(g_{m_1} + g_{ds_1} + g_{ds_3} + \\ + s(C_{db_1} + C_{gs_1} + C_{db_3} + C_{gd_3})) \\ V_x(g_{m_1} + g_{ds_1} + sC_{db_1}) = V_{ip}(g_{m_1} - sC_{gd_1}) - V_{in}(g_{m_2} + sC_{gs_2}) + \\ + V_{out}(g_{m_2} + g_{mb_2} + g_{ds_1} + g_{ds_2} + g_{ds_4} + s(C_{db_1} + C_{gd_1} + C_{gs_2} + \\ + C_{sb_2} + C_{db_4} + C_{gd_4} + C_L)) \end{cases} \quad (5.31)$$

Combining the equations in Eq. 5.31 a transfer function with two poles and two zeros is obtained. Due to its size the transfer function is not presented here.

5.5.3   Simulation Results

The two amplifiers presented in the previous sections were simulated in order to determine which one offers the best performance. Figure 5.40 shows the gain of the amplifiers in function of the DC voltage. From the figure it can be concluded that the amplifiers (normal version and complementary version) have low linearity around half the supply voltage (0.6 V) although when the supply voltage is decreased to 0.9 V, the linearity is improved at half the supply voltage (0.45 V), fact that can be verified by the distortion values obtained from simulation (Fig. 5.40). The linearity is improved, at the cost of losing gain, using source degeneration, which will be discussed ahead.

Assuming that the clock frequency of the filter is 10 MHz, this means that the current used in the amplifier (single-ended) should be chosen so that the GBW (Eq. 5.21) is higher than 22 MHz. For both versions the current used in the simulations was 100 μA.

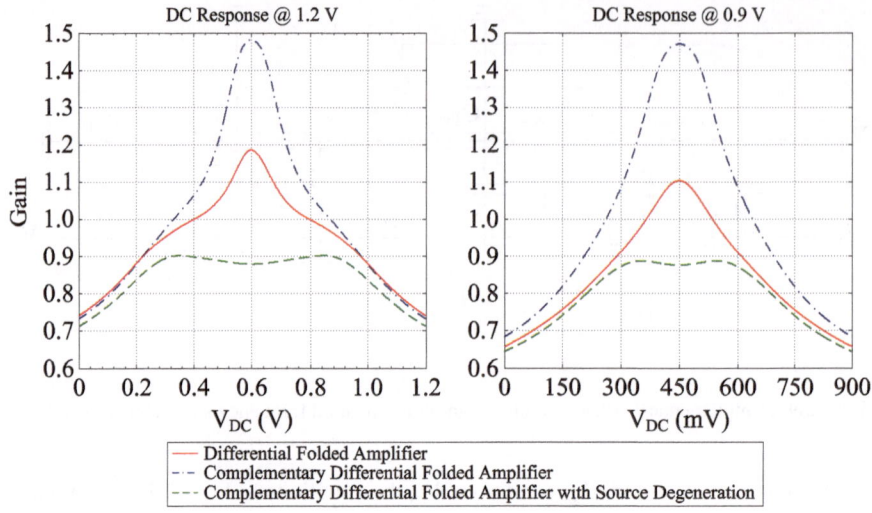

Fig. 5.40 Relation between the gain of the circuits and the DC voltage

Table 5.8 Simulated results for the amplifiers with a common mode voltage of 0.6 V: (a) Differential folded amplifier (b) Complementary differential folded amplifier

Version	Supply (V)	Power (μW)	Gain	GBW (MHz)	THD (dB)	Noise (μV)	FOM
(a)	1.2	239	1.186	62.3	−40.01	34.6	1106
	0.9	171	1.103	63.5	−43.84	32.2	1575
(b)	1.2	259	1.481	57.7	−39.53	39.8	947
	0.9	183	1.470	54.9	−45.89	40.3	1275

The results obtained from simulating the amplifiers with a common mode voltage of 0.6 V, input signal with an amplitude of Supply/24, and a load capacitance of 4.25 pF are shown in Table 5.8. The results show that the THD of the amplifiers is relatively high (between −40 and −46 dB).

One of the reasons that contributes to the high distortion is the circuits common mode voltage value, which was already described for Sect. 5.4.4 buffer. Another reason that increases this amplifiers distortion is the fact that, in differential config-uration, transistors M_1 have their sources connected between each other, forming a differential pair. The problem with the linearity of the differential pair comes from not having a V_{GS} large enough since the linear input range is small, which degrades the amplifiers linearity.

A possible way to improve the amplifiers distortion, besides adjusting the common mode voltage, is to use a linearization technique. A possible technique [14, 15 16] involves using source degeneration in the differential pair, using two MOS transistors operating in the triode region (M_5 and M_6). The complementary differential folded amplifier using source degeneration is shown in Fig. 5.41. Using this technique it is

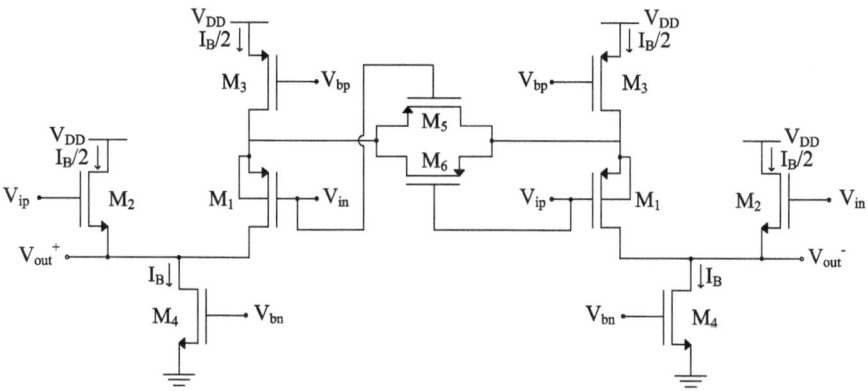

Fig. 5.41 Complementary differential folded amplifier with source degeneration using MOS transistors

Table 5.9 Simulated results for the complementary differential folded amplifier with source degeneration

Supply (V)	Power (μW)	Gain	GBW (MHz)	THD (dB)	Noise (μV)	FOM
1.2	259	0.879	58.2	-63.37	40.0	955
0.9	183	0.876	55.3	-61.87	40.4	1285

possible to enhance the amplifiers linearity since transistors operating in the triode region exhibit higher linearity than transistors operating in the saturation region [15].

The results obtained from simulating the complementary differential folded amplifier with source degeneration are shown in Table 5.9. The results show that by using source degeneration in the differential pair it is possible to improve the distortion values in over 10 dB. The distortion value could be improved further if the width of transistors M_5 and M_6 were to be decreased further, which would increase these transistors resistance and their linearity. The downside of doing this would be that the gain would decrease, making it more difficult to find a viable solution for the capacitors in the SC filters.

The transistor sizes used for the simulations in this section are shown in Table 5.10. The sizes used in the complementary differential folded amplifier with source degeneration are identical to the ones used in the complementary differential folded amplifier.

Table 5.10 Transistor parameters: (a) Differential folded amplifier (DFA) with supply @ 1.2 V (b) DFA with supply @ 0.9 V (c) Complementary differential folded amplifier (CDFA) with supply @ 1.2 V (d) CDFA with supply @ 0.9 V

		$W(\mu m)$	$L(\mu m)$	$I_D(\mu A)$	$g_m(mS)$	$g_{ds}(\mu S)$	$V_{DSAT}(mV)$
(a)	M_1	6.00	0.12	23.97	0.51	38.22	54.89
	M_2	41.32	0.12	−75.76	1.53	130.78	−76.30
	M_3	2.40	0.12	23.97	0.43	37.84	71.71
	M_{3B}	0.24	0.12	5	0.08	6.06	89.34
	M_4	72.00	0.36	−99.73	1.55	19.34	−109.27
	M_{4B}	7.20	0.36	−10	0.16	1.86	−109.24
(b)	M_1	6.00	0.12	18.79	0.40	29.59	52.30
	M_2	41.32	0.12	−76.33	1.53	131.69	−77.62
	M_3	2.40	0.12	18.79	0.35	48.03	69.62
	M_{3B}	0.24	0.12	5	0.08	6.06	89.34
	M_4	72.00	0.36	−95.12	1.46	60.00	−108.85
	M_{4B}	7.20	0.36	−10	0.16	1.86	−109.24
(c)	M_1	90.00	0.36	−50.12	1.00	9.60	−77.14
	M_2	60.00	0.36	57.78	1.40	19.72	57.68
	M_3	49.98	0.36	−50.12	0.86	10.34	−96.43
	M_{3B}	5.00	0.36	−5	0.09	0.98	−96.25
	M_4	30.00	0.36	107.90	2.06	41.56	83.97
	M_{4B}	3.00	0.36	10	0.19	3.82	82.19
	$M_{5,6}$	0.50	0.36	0	0	1.97	−55.11
(d)	M_1	90.00	0.36	−47.86	0.97	9.23	−76.02
	M_2	60.00	0.36	53.76	1.30	19.35	56.97
	M_3	49.98	0.36	−47.86	0.81	31.81	−96.08
	M_{3B}	5.00	0.36	−5.00	0.09	0.98	−96.25
	M_4	30.00	0.36	101.62	1.94	58.94	83.96
	M_{4B}	3.00	0.36	10.00	0.19	3.82	82.19
	$M_{5,6}$	0.50	0.36	0	0	3.09	−61.50

Chapter 6
Switched Capacitor Filter Implementation

Abstract In this chapter the filter circuits will be simulated taking into account the non-ideal effects described in fifth chapter. Due to these effects the band-pass filter will be altered, adding a new capacitor to the circuit. This is done in order to take into account the input capacitance of the low gain amplifier and facilitate the compensation of this capacitance. The simulated circuits include second-order low-pass filters, two operating at 1.2 V using the buffers presented in the fifth chapter, one operating at 0.9 V, and a sixth-order filter obtained from cascading three biquadratic sections operating at 1.2 V. The band-pass filter circuit was also simulated for a second-order filter, and a fourth-order filter obtained from cascading two biquadratic sections, all operating at 1.2 V.

6.1 Second-Order Low-Pass SC Filter

The simulation results of the second-order low-pass SC filter using real components will be presented in this section. The results of the simulations will be presented when all components are ideal, when the switches are real and the remaining components ideal, when the buffers/amplifiers are real and the remaining components ideal, and when all components are real in order to determine the influence of each real component in the overall performance of the filter.

When replacing the ideal buffer with a real buffer the value of capacitance C_2 must be recalculated in order to compensate the input capacitance of the real buffer. An approximate value for the input capacitance can be obtained from the DC simulation of the circuit. When using real switches instead of ideal, the filters capacitance values must also be recalculated in order to compensate the switches parasitic capacitances C_d and C_s. An approximate value for these capacitances can also be obtained from the DC simulation. The single-ended version of the low-pass filter considering the switches parasitic capacitances and the buffers input capacitance is shown in Fig. 6.1.

© Springer International Publishing Switzerland 2015

63

H. A. de A. Serra, N. Paulino, *Design of Switched-Capacitor Filter Circuits using Low Gain Amplifiers,* SpringerBriefs in Electrical and Computer Engineering, DOI 10.1007/978-3-319-11791-1_6

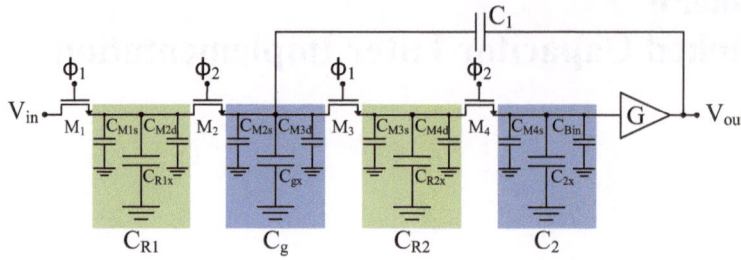

Fig. 6.1 Low-pass SC filter considering parasitic capacitances

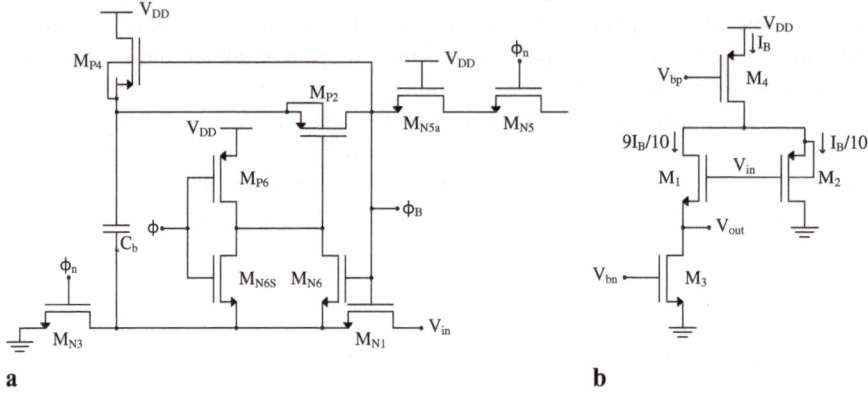

Fig. 6.2 Clock boost (**a**) and buffer (**b**) circuits used in this section

The compensated value of the capacitors can be obtained using the equations in Eq. 6.1.

$$\begin{cases} C_{gx} = C_g - C_{M2s} - C_{M3d} \\ C_{2x} = C_2 - C_{M4s} - C_{Bin} \\ C_{R1x} = C_{R1} - C_{M1s} - C_{M2d} \\ C_{R2x} = C_{R2} - C_{M3s} - C_{M4d} \end{cases} \tag{6.1}$$

6.1.1 Filter using Clock Boost and Complementary Source Follower with g_{ds} Compensation

In this section the simulation results for the second-order low-pass filter using the clock boost presented in Sect. 5.2 and buffer of Sect. 5.4.2 (shown here again for convenience) are presented when operating at supply voltage of 1.2 V.

The capacitance values used (in differential configuration) in the simulations and the distortion results of the filter are shown in Table 6.1. In the first three simulations

Table 6.1 Filter capacitances and distortion results: (a) Using ideal components (b) Using real buffers (c) Using real switches (d) Using real components (⋆ single-ended capacitor)

	$C_{R1}(fF)$	$C_{R2}(fF)$	$C_g^{\star}(fF)$	$C_1^{\star}(pF)$	$C_2(fF)$	THD (dB)
(a)	106.72	70	400	4	760.64	−96.35
(b)	106.72	70	400	4	725	−54.02
(c)	106.68	69.95	399.91	4	760.59	−74.95
(d)	106.66	69.98	399.82	4	724.79	−66.84

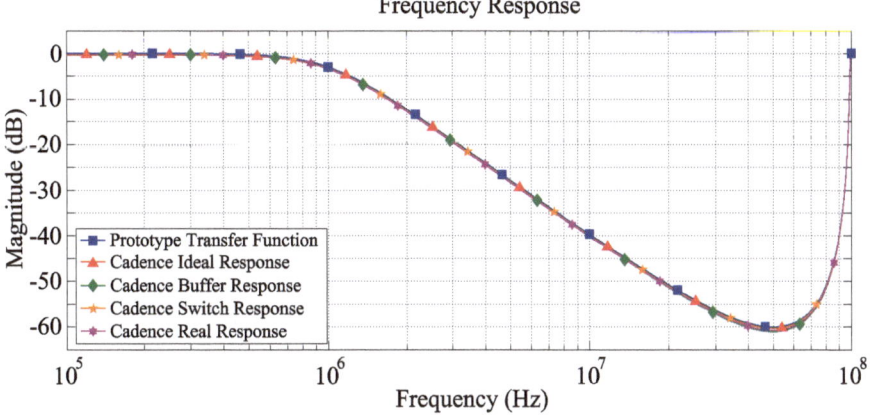

Fig. 6.3 Frequency responses of the second-order filter

(a, b, and c) the filter was simulated using ideal capacitors. In the last simulation (d) the filter was simulated using real components, including real capacitors (MIM-CAPS_MML130E). The value that was assumed for the switches source and drain capacitances was 290 pF/m, and the gain of the buffer was 0.97.

The frequency responses are shown Fig. 6.3. The attenuations at the cutoff frequency for each simulation were 3.00 dB, 3.28 dB, 3.44 dB, 3.38 dB, and 3.56 dB, respectively.

6.1.2   Filter Using Low-Voltage Clock Boost and Voltage Combiner at 1.2 V

In this section the simulation results for the second-order low-pass filter using the low-voltage clock boost presented in Sect. 5.3 and low-voltage amplifier of Sect. 5.5.2 (shown here again for convenience) with source degeneration (Fig. 5.41) are presented when operating at supply voltage of 1.2 V.

Table 6.2 shows the values of the capacitors used in the simulations and the filters distortion. The value that was assumed for the switches source and drain capacitances was 290 pF/m, and the gain of the amplifier was 0.879.

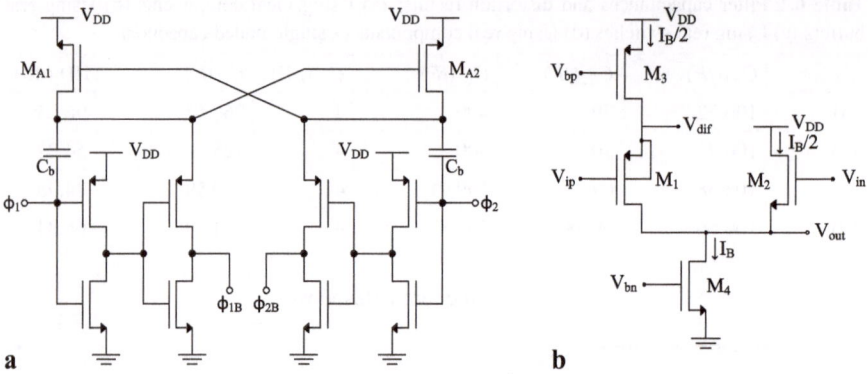

Fig. 6.4 Clock boost (**a**) and amplifier (**b**) circuits used in this section

Table 6.2 Filter capacitances and distortion results: (a) Using ideal components (b) Using real amplifiers (c) Using real switches (d) Using real components ($\star$ single-ended capacitor)

	C_{R1}(fF)	C_{R2}(fF)	C_g*(fF)	C_1*(pF)	C_2(fF)	THD (dB)
(a)	108.44	60	500	4.25	612.44	−105.77
(b)	108.44	60	500	4.25	560	−62.20
(c)	108.39	59.95	499.91	4.25	612.42	−75.41
(d)	108.43	60	500.03	4.25	560.02	−62.01

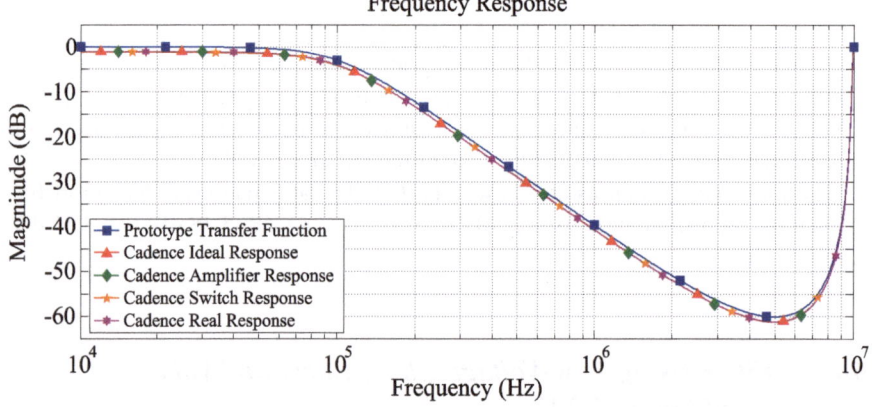

Fig. 6.5 Frequency responses of the second-order filter at 1.2 V

The frequency responses are shown Fig. 6.5. The attenuations at the cutoff frequency for each simulation were 3.00, 4.07, 4.08, 4.17, and 4.07 dB, respectively. The attenuation before the cutoff frequency is more visible in this case due to the amplifier having an attenuation of 1.12 dB.

Table 6.3 Filter capacitances and distortion results: (a) Using ideal components (b) Using real amplifiers (c) Using real switches (d) Using real components (⋆ single-ended capacitor)

	$C_{R1}(fF)$	$C_{R2}(fF)$	$C_g{}^\star(fF)$	$C_1{}^\star(pF)$	$C_2(fF)$	THD (dB)
(a)	108.08	60	500	4.25	610.34	−102.18
(b)	108.08	60	500	4.25	555	−61.32
(c)	108.04	59.95	499.91	4.25	610.32	−76.38
(d)	108	60	500.03	4.25	554.83	−64.72

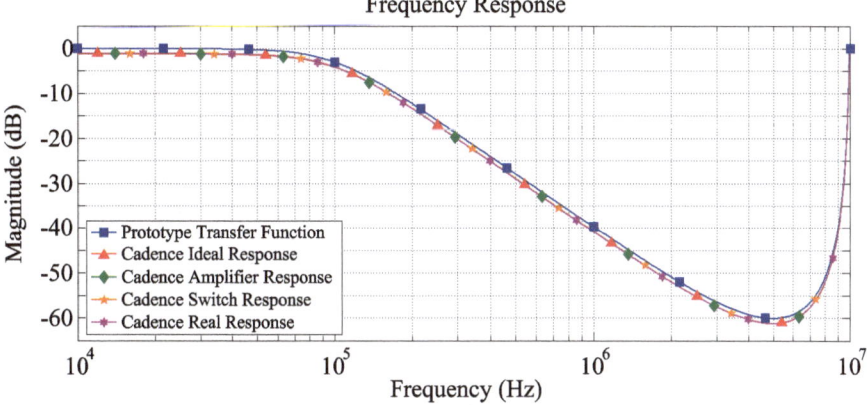

Fig. 6.6 Frequency responses of the second-order filter at 0.9 V

6.1.3   Filter Using Low-Voltage Clock Boost and Voltage Combiner at 0.9 V

In this section the filter is simulated using the same clock boost circuit and amplifier from the previous section (Fig. 6.4) but with supply voltage of 0.9 V instead of 1.2 V.

Table 6.3 shows the values of the capacitors used in the simulations and the filters distortion. The value that was assumed for the switches source and drain capacitances was 29 pF/m, and the gain of the amplifier was 0.876.

The frequency responses are shown Fig. 6.6. The attenuations at the cutoff frequency for each simulation were 3.00, 4.08, 4.12, 4.16, and 4.10 dB, respectively.

6.2   Sixth-Order Low-Pass SC Filter

The sixth-order low-pass filter was implemented by cascading three second-order low-pass filters with different quality factors as described in Sect. 3.3. The results of the simulations are presented in the next sections when all components are ideal, when the switches are real and the remaining components ideal, when the buffers/amplifiers are real and the remaining components ideal, and when all components are real.

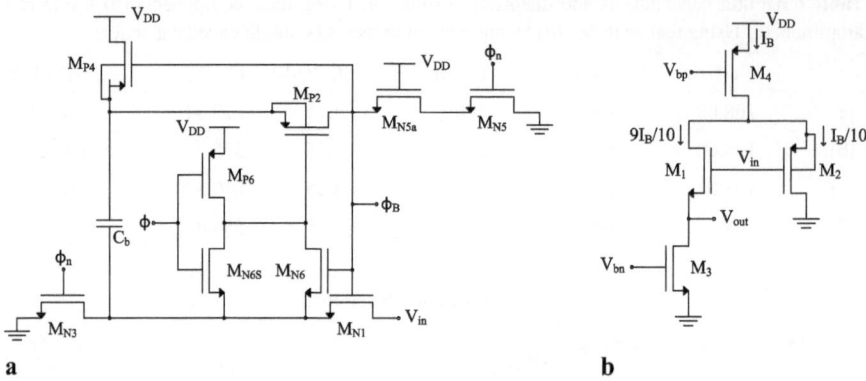

Fig. 6.7 Clock boost (**a**) and buffer (**b**) circuits used in this section

Table 6.4 Filter capacitances and distortion results: (a) Using ideal components (b) Using real buffers (c) Using real switches (d) Using real components ($\star$ single-ended capacitor)

	$C_{R1}(fF)$	$C_{R2}(fF)$	$C_g{}^\star(fF)$	$C_1{}^\star(pF)$	$C_2(fF)$	THD (dB)
(a)	194.01	70	400	4	1407.60	−95.00
(b)	194.01	70	400	4	1360	−55.36
(c)	193.92	69.91	399.81	4	1407.51	−70.04
(d)	193.87	69.82	399.67	4	1359.53	−67.67

The filter sections will alternate between the complementary and normal version of the buffers. The low quality factor and the high quality factor filters will use the complementary version while the medium quality factor filter will use the normal version. This is done in order to maintain the common mode voltage as far away as possible from V_{DD} and 0 V to obtain better distortion values.

6.2.1 Section with Low Quality Factor

Table 6.4 shows the values of the capacitors used in the simulations and the filters distortion. The value that was assumed for the switches source and drain capacitances was 291 pF/m (W = 320 nm), and the gain of the buffer was 0.97.

The frequency responses are shown Fig. 6.8. The attenuations at the cutoff frequency for each simulation were 5.71, 5.98, 6.08, 6.08, and 6.27 dB, respectively.

6.2.2 Section with Medium Quality Factor

Table 6.5 shows the values of the capacitors used in the simulations and the filters distortion. The value that was assumed for the switches source and drain capacitances was 291 pF/m (W = 160 nm), and the gain of the buffer was 0.98.

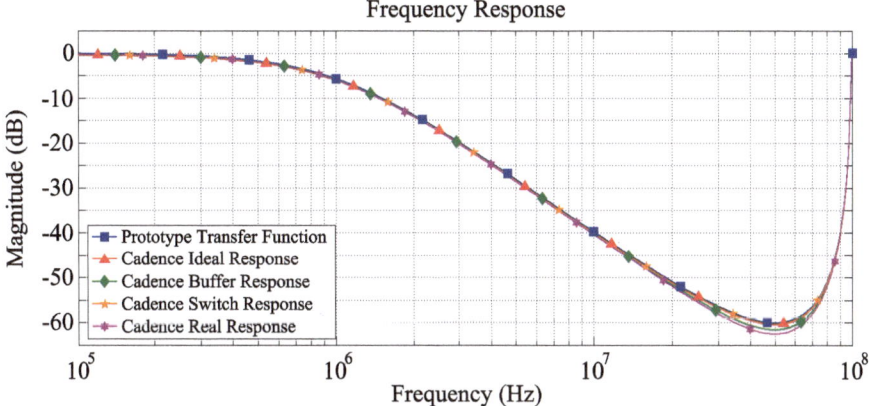

Fig. 6.8 Frequency responses of the low quality factor filter

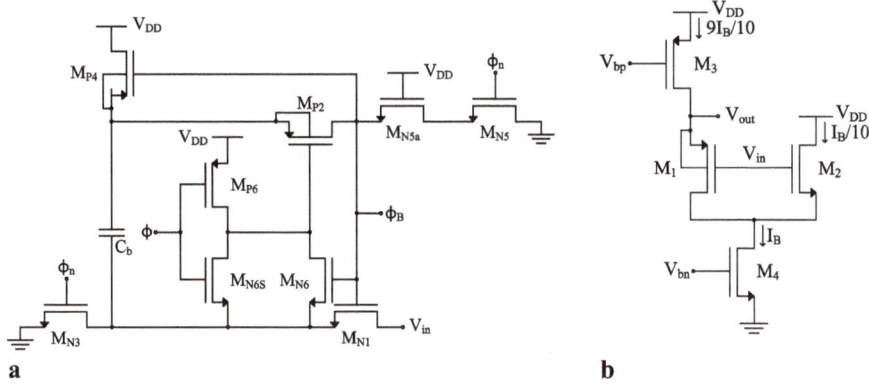

Fig. 6.9 Clock boost (**a**) and buffer (**b**) circuits used in this section

Table 6.5 Filter capacitances and distortion results: (a) Using ideal components (b) Using real buffers (c) Using real switches (d) Using real components (⋆ single-ended capacitor)

	$C_{R1}(fF)$	$C_{R2}(fF)$	$C_g{}^\star(fF)$	$C_1{}^\star(pF)$	$C_2(fF)$	THD (dB)
(a)	109.36	70	400	4	780.17	−97.49
(b)	109.36	70	400	4	730	−54.10
(c)	109.31	69.95	399.91	4	780.12	−72.41
(d)	109.36	69.94	399.87	4	729.76	−54.93

The frequency responses are shown Fig. 6.10. The attenuations at the cutoff frequency for each simulation were 3.00, 3.18, 3.36, 3.26, and 3.50 dB, respectively.

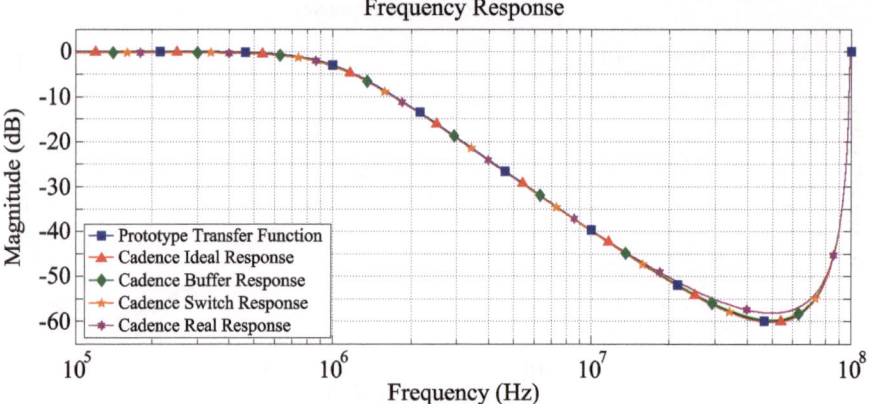

Fig. 6.10 Frequency responses of the medium quality factor filter

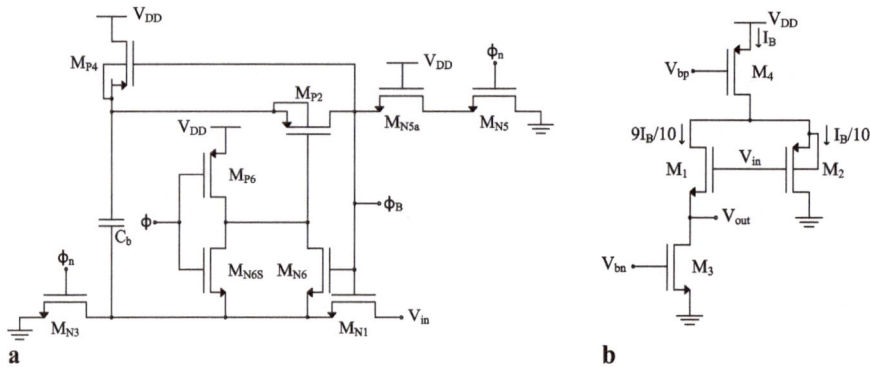

Fig. 6.11 Clock boost (**a**) and buffer (**b**) circuits used in this section

Table 6.6 Filter capacitances and distortion results: (a) Using ideal components (b) Using real buffers (c) Using real switches (d) Using real components ($\star$ single-ended capacitor)

	$C_{R1}(fF)$	$C_{R2}(fF)$	$C_g{}^\star(fF)$	$C_1{}^\star(pF)$	$C_2(fF)$	THD (dB)
(a)	159.14	50	50	16	198.80	−95.63
(b)	159.14	50	50	16	30	−51.47
(c)	159.09	49.95	49.91	16	198.76	−63.38
(d)	158.96	49.96	49.90	16	29.85	−56.81

6.2.3   Section with High Quality Factor

Table 6.6 shows the values of the capacitors used in the simulations and the filters distortion. The value that was assumed for the switches source and drain capacitances was 291 pF/m (W = 160 nm), and the gain of the buffer was 0.97.

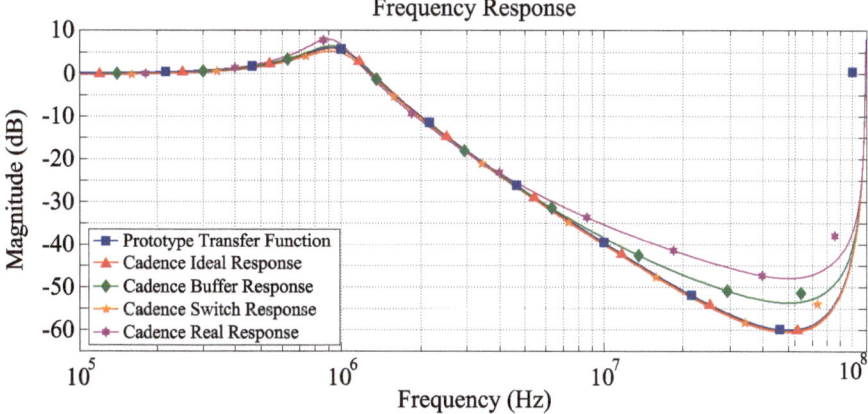

Fig. 6.12 Frequency responses of the high quality factor filter

This section has the poles of the transfer function close to the unit circle which makes the sections response sensitive to component value variations. This fact can be seen in the difference between the ideal and real frequency responses (Fig. 6.12) making this filter section the one that presents worse results. Other factors that contributed for the poor performance were the buffer used and the capacitance values. Since the buffer needs to charge a 16 pF capacitor, the buffers current and size had to be increased by a factor of four from the ones presented in Table 5.5(d). Doing this increased the input capacitance of the buffer substantially. Due to the gain of the buffer and the high quality factor of this filter section, the values obtained for the capacitors were small (except for C_1). All these factors made the filters response more susceptible to non-linearities.

The frequency responses are shown Fig. 6.12. The gain at the cutoff frequency for each simulation were 5.73, 5.46, 5.55, 4.79, and 6.05 dB, respectively.

6.2.4   Cascaded Sections

For the simulation of the cascaded sections the value of the capacitors were slightly altered to compensate for the buffers gain variation due to the common mode voltage being different from the previous simulations. The distortions of the complete filter were −95.98, −46.74, −55.46 and −53.98 dB, respectively.

The frequency responses are shown Fig. 6.13. The attenuations at the cutoff frequency for each simulation were 2.99, 3.82, 3.28, 4.50, and 3.04 dB, respectively.

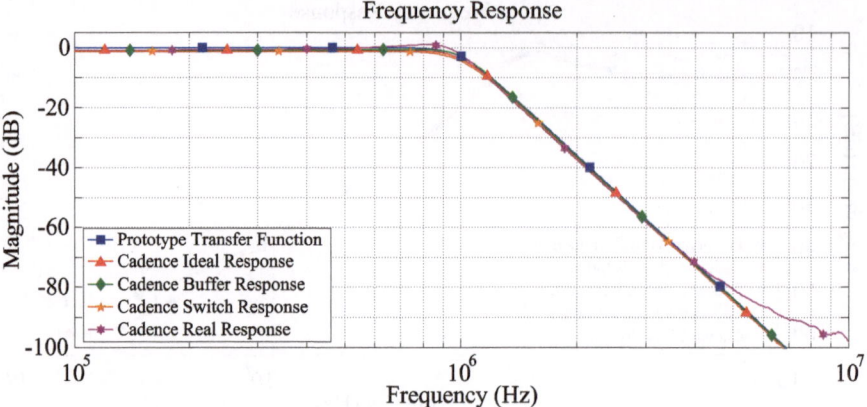

Fig. 6.13 Frequency responses of the sixth-order filter

6.3 Second-Order Band-Pass SC Filter

In the simulations of the second-order band-pass SC filter the ground voltage of the filter was altered from 0 V to $V_{DD}/2$. This was done since during clock phase ϕ_2 capacitor C_{R2} is connected to ground, discharging and moving the common mode voltage of the filter to 0 V. By switching the ground voltage to $V_{DD}/2$ the common mode voltage at the entrance to the amplifier remains at $V_{DD}/2$, ensuring that it works as intended. For the sizing of this filter the parasitic capacitances also have to be taken into account. Figure 6.14 shows the single-ended version of the band-pass filter considering the switches parasitic capacitances and the amplifiers input capacitance. In order to facilitate the compensation of the parasitic capacitances, an additional capacitance (C_B) was added to the filters circuit at the entrance of the amplifier. Equation 6.2 shows the charge equations of the new filter circuit.

$$\begin{cases} A \longrightarrow & V_{in}[n-1]C_{R1} + V_{c1}[n-1](C_1+C_2) + V_{out}[n-1](C_{Rf}-C_2/G) = \\ & = V_{c1}[n-0.5](C_{R1}+C_1+C_{Rf}+C_2) - V_{out}[n-0.5]C_2/G \\ B \longrightarrow & V_{out}[n-1](C_2+C_B)/G - V_{c1}[n-1]C_2 = \\ & = V_{out}[n-0.5](C_2+C_B)/G - V_{c1}[n-0.5]C_2 \\ C \longrightarrow & V_{c1}[n-0.5](C_1+C_2) - V_{out}[n-0.5]C_2/G = \\ & = V_{c1}[n](C_1+C_2) - V_{out}[n]C_2/G \\ D \longrightarrow & V_{out}[n-0.5](C_2+C_B)/G - V_{c1}[n-0.5]C_2 = \\ & = V_{out}[n](C_2+C_{R2}+C_B)/G - V_{c1}[n]C_2 \end{cases}$$

$$(6.2)$$

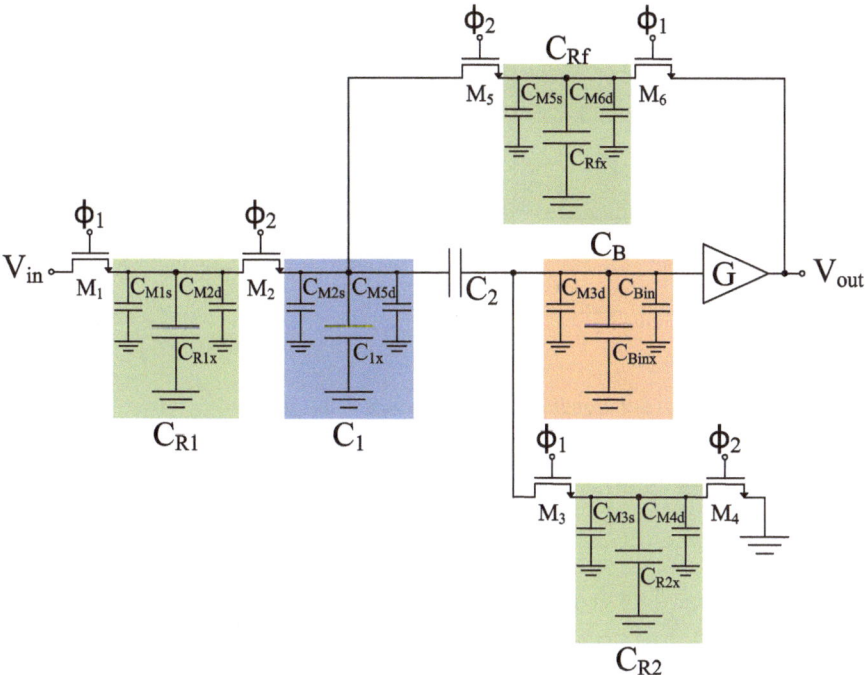

Fig. 6.14 Band-pass SC filter considering parasitic capacitances [17]

The compensated value of the capacitors can be obtained using the equations in Eq. 6.3.

$$\begin{cases} C_{R1x} = C_{R1} - C_{M1s} - C_{M2d} \\ C_{R2x} = C_{R2} - C_{M3s} - C_{M4d} \\ C_{Rfx} = C_{Rf} - C_{M5s} - C_{M6d} \\ C_{1x} \ \ = C_1 - C_{M2s} - C_{M5d} \\ C_{Binx} = C_B - C_{Bin} - C_{M3d} \end{cases} \tag{6.3}$$

The capacitance values used (in differential configuration) in the simulations are shown in Table 6.7. In the first two simulations (a and b) the filter was simulated using ideal capacitors. In the last simulation (c) the filter was simulated using real components, including real capacitors (MIMCAPS_MML130E). The value that was assumed for the switches source and drain capacitances was 301 and 321 pF/m, respectively.

The frequency responses are shown Fig. 6.16. The attenuations at the cutoff frequency for each simulation were [2.16, 1.87], [2.09, 2.00], and [1.86, 1.79] dB, respectively.

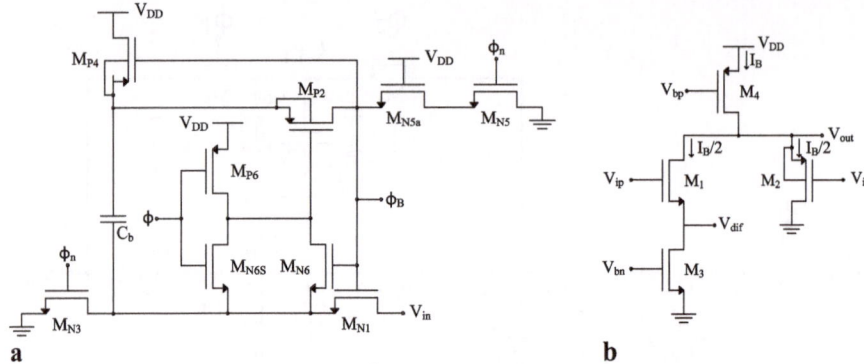

Fig. 6.15 Clock boost (**a**) and amplifier (**b**) circuits used in this section

Table 6.7 Filter capacitances: (a) Using ideal components (b) Using real switches (c) Using real components ($\star$ single-ended capacitor)

	$C_{R1}(fF)$	$C_{R2}(fF)$	$C_{Rf}(fF)$	$C_1(fF)$	$C_2^\star(pF)$	$C_B^\star(fF)$
(a)	53.09	53.16	312.55	1500.00	4	500.00
(b)	52.94	53.00	312.39	1499.84	4	499.84
(c)	53.00	53.00	312.30	1500.10	4	389.14

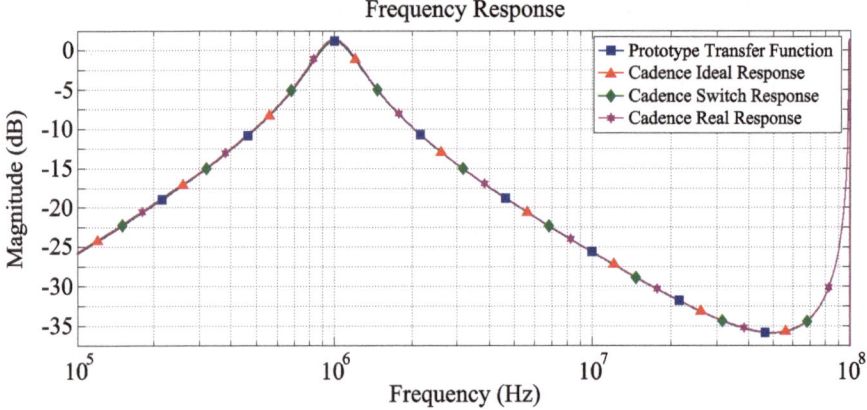

Fig. 6.16 Frequency responses of the filter

6.4 Fourth-Order Band-Pass SC Filter

The fourth-order band-pass filter was implemented by cascading two second-order band-pass filters as described in Sect. 4.3. The results of the simulations are presented in the next sections when all components are ideal, when the switches are real and the remaining components ideal, when the buffers/amplifiers are real and the remaining components ideal, and when all components are real. Like in the sixth-order low-pass

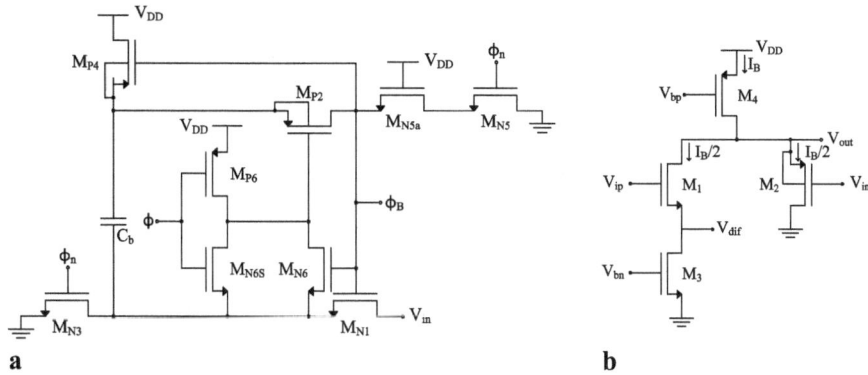

Fig. 6.17 Clock boost (**a**) and amplifier (**b**) circuits used in this section

filter, each section uses a specific amplifier to maintain the common mode voltage. The first section will use the normal version of the amplifier while the second uses the complementary version.

6.4.1 First Filter Section

Table 6.8 shows the values of the capacitors used in the simulations and the filters distortion. The value that was assumed for the switches source and drain capacitances was 301 and 321 pF/m (W = 500 nm), respectively, and the gain of the amplifier was 1.22. The current used by the amplifier was 1 mA.

Table 6.8 Filter capacitances: (a) Using ideal components (b) Using real switches (c) Using real components (⋆ single-ended capacitor)

	$C_{R1}(fF)$	$C_{R2}(fF)$	$C_{Rf}(fF)$	$C_1(fF)$	$C_2^\star(pF)$	$C_B^\star(fF)$
(a)	59.80	21.03	498.55	500	4	500
(b)	59.64	20.88	498.40	499.84	4	499.84
(c)	59.59	20.86	498.43	500.01	4	359.93

The frequency responses obtained from the impulse responses are shown Fig. 6.18. The attenuations at the cutoff frequency for each simulation were [1.09, 0.82] dB, [0.83, 0.73] dB, [0.67, 0.55] dB, and [0.25, 0.49] dB, respectively.

6.4.2 Second Filter Section

Table 6.9 shows the values of the capacitors used in the simulations and the filters distortion. The value that was assumed for the switches source and drain capacitances

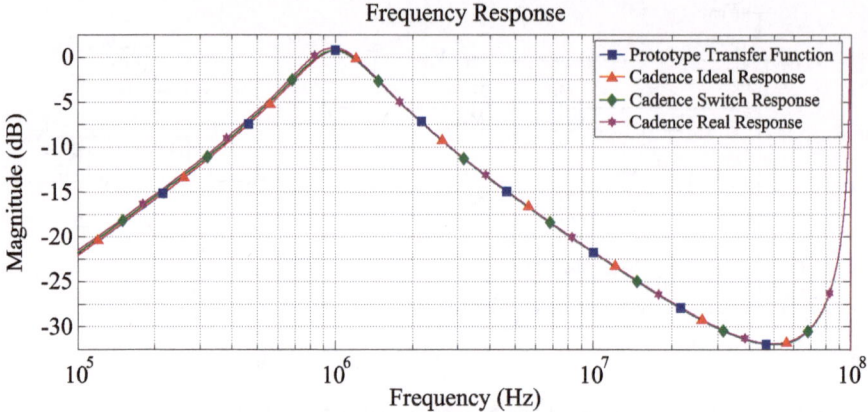

Fig. 6.18 Frequency responses of the first section of the filter

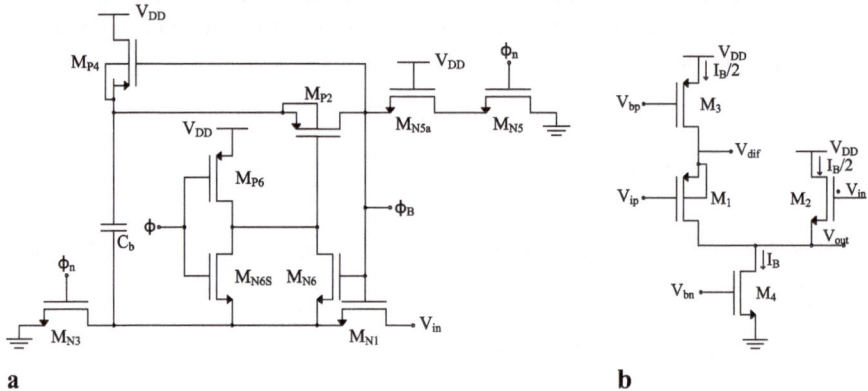

Fig. 6.19 Clock boost (**a**) and amplifier (**b**) circuits used in this section

Table 6.9 Filter capacitances: (a) Using ideal components (b) Using real switches (c) Using real components (⋆ single-ended capacitor)

	$C_{R1}(fF)$	$C_{R2}(fF)$	$C_{Rf}(fF)$	$C_1(fF)$	$C_2^\star(pF)$	$C_B^\star(fF)$
(a)	51.59	25.43	359.85	500	4	500
(b)	51.44	25.28	359.69	499.84	4	499.84
(c)	51.47	25.28	359.72	499.69	4	350.13

was 301 and 321 pF/m ($W = 500$ nm), respectively, and the gain of the amplifier was 1.26. The current used by the amplifier was 1 mA.

The frequency responses are shown Fig. 6.20. The attenuations at the cutoff frequency for each simulation were [1.09, 0.82] dB, [0.77, 0.50] dB, [0.65, 0.61] dB, and [0.89, 0.78] dB, respectively.

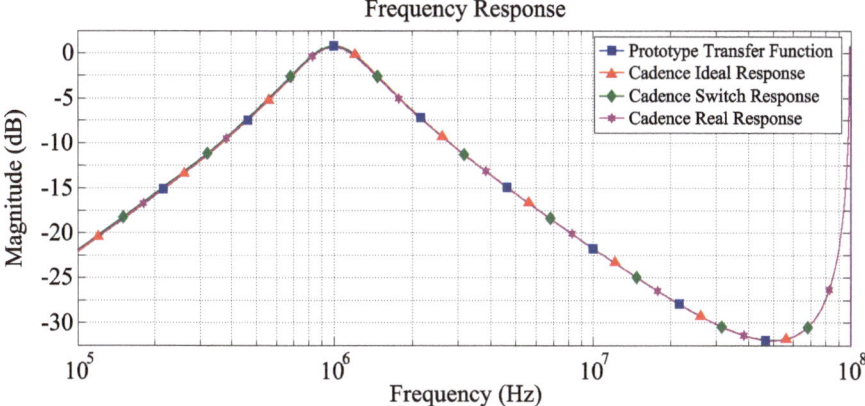

Fig. 6.20 Frequency responses of the second section of the filter

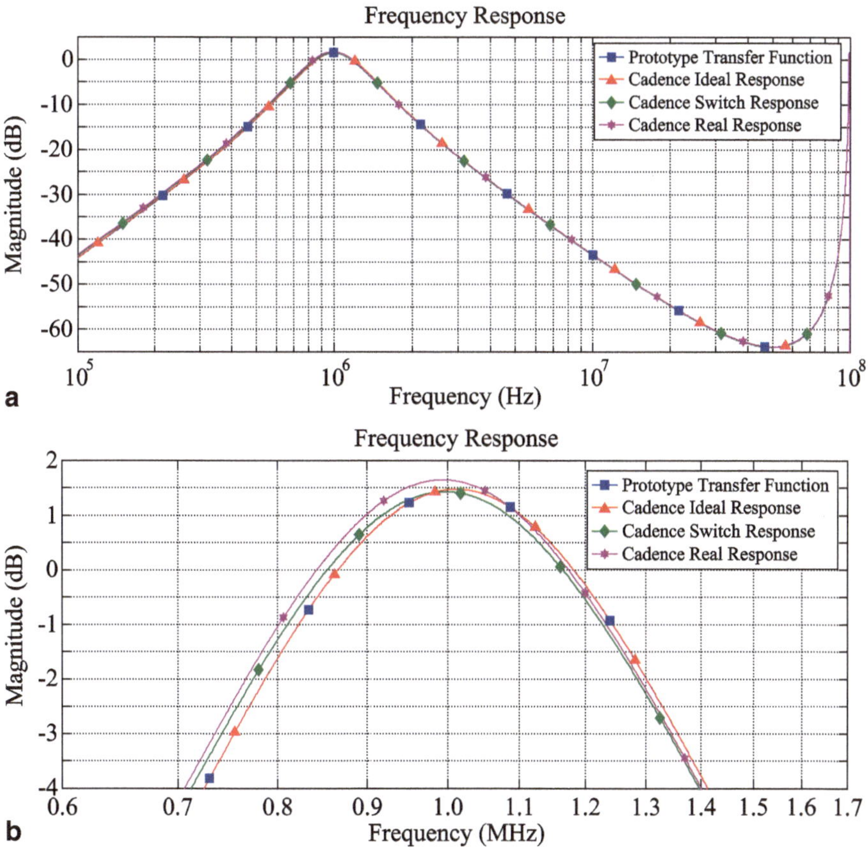

a

b

Fig. 6.21 Frequency responses of the fourth-order filter

6.4.3 Cascaded Sections

The frequency responses are shown Fig. 6.21. The attenuations at the cutoff frequency for each simulation were [2.18, 1.64] dB, [1.88, 1.33] dB, [1.55, 1.64] dB, and [1.37, 1.60] dB, respectively.

Chapter 7
Conclusion

Abstract This chapter summarizes the work performed to design low-pass and band-pass Sallen-Key SC filters, using low gain amplifiers, in modern nm CMOS technologies.

The objective of the work presented in this book was to convert the low-pass and band-pass Sallen-Key topologies to SC circuits and adapt them to work in modern nm CMOS technologies. To do this, different activities were preformed and several results were obtained, which are summarized next.

In the third chapter, the low-pass Sallen-Key filter topology, from an ideal perspective, was presented along with the filters transfer function. This filter was then converted into a SC filter by replacing the resistors with parallel SC networks. During the simulation phase it was seen that the filter would lose charge during operation due to the feedback capacitor floating between phases. To resolve this issue, a capacitor connected to ground was added at the floating node. With this alteration in the filter the design equations were obtained and the filter was designed using the Butterworth prototype transfer function and the forward Euler transform. During this step it was found that there were not enough variables to find a viable solution for both the numerator and the denominator of the prototype transfer function. Due to this the filter was designed using only the Butterworth denominator coefficients and the filters DC gain became dependent on the buffers attenuation.

Once the capacitance values and gain were determined, the filters impulse response was simulated in Cadence using ideal components to determine that the filter was working as intended. The differential configuration of the filter was then obtained and compared with the single-ended configuration in order to eliminate the even harmonics and reduce the noise level due to charge injection from the switches. In this chapter a sixth-order low-pass filter obtained from cascading three biquadratic sections was also analyzed and simulated from an ideal standpoint.

In the fourth chapter, the band-pass Sallen-Key filter topology, from an ideal perspective, was presented and the same steps from the previous chapter were taken. This topology, however, did not have problems of floating capacitors, but due to capacitor C_{R2} discharging to ground during clock phase ϕ_2 the common mode voltage of the filter was lost. To avoid losing the common mode voltage, the capacitor was connected to $V_{DD}/2$ during clock phase ϕ_2. Unlike the low-pass version, the band-pass filter was designed using both the numerator and the denominator of the prototype transfer function. In the simulation using real components, the filter had

H. A. de A. Serra, N. Paulino, *Design of Switched-Capacitor Filter Circuits using Low Gain Amplifiers,* SpringerBriefs in Electrical and Computer Engineering, DOI 10.1007/978-3-319-11791-1_7

a small gain over the prototype frequency response. A fourth-order band-pass filter obtained from cascading two biquadratic sections was also analyzed and simulated from an ideal standpoint.

In the fifth chapter, the non-ideal effects of using real components was studied. Three types of switches were simulated in a first-order SC filter, to determine which one offered the best performance in terms of harmonic distortion. From the simulations, it was determined that increasing the width of the switches reduces the non-linear effects introduced by the switches resistance, but increasing the width of the switches increases the non-linear effects of the parasitic capacitances, increasing the difficulty in compensating them during the design phase. Overall, it was determined that a NMOS switch driven by a boosted clock signal offers the best performance when narrow switches are used. Two clock boost circuits were presented and simulated, one for regular 1.2 V operation and one for operation at both 1.2 and 0.9 V. For this second clock boost circuit it was determined that the minimum supply voltage at which this circuit correctly operates depends on the clock frequency. If the clock frequency is increased beyond the circuit bandwidth the phases will overlap and cannot be used in the SC filters.

The non-ideal effects of buffers were also analyzed in the fifth chapter. Two buffers were analyzed (DC gain, thermal noise referred to the input, and transfer function) in their normal and in their complementary configuration. This was done to allow the cascading of biquadratic sections and maintain the common mode voltage away from the ground and supply voltages, since signals begin to saturate close to these voltages increasing the harmonic distortion they introduce in the filter. In the complementary configuration of the source follower with g_{ds} compensation, which is used in the low-pass filters, it was seen that the gain of the circuit was lower than the normal configuration due to the body effect in one of the transistors. To compensate the body effect, RF transistor models were used in both configurations since both NMOS and PMOS devices can connect their bulks and sources together, when using these models, eliminating the body effect and improving the buffers gain. To improve the harmonic distortion introduced by these buffers, the common mode voltage was adjusted from $V_{DD}/2$ to $(V_{DD} - V_{GS_1})/2$ so that the common mode voltage value of both the input and output signal stay as far away as possible from the supply voltage and ground. Using this technique the harmonic distortion improved in over 10 dB. Two configurations of the fully-differential voltage combiner were also analyzed. This amplifier is used in the band-pass filter since it has a gain higher than 1 and is also used in the low-voltage low-pass filter with source degeneration to improve the differential pairs linearity at the cost of losing gain. Using the source degeneration technique it was possible to improve the amplifiers harmonic distortion by over 10 dB.

In the seventh chapter, several filters were simulated: when all components are ideal, when the switches are real and the remaining components ideal, when the buffers/amplifiers are real and the remaining components ideal, and when all components are real in order to determine the influence of each non-ideal component in the frequency response of the filter. From the simulation it was concluded that the buffer/amplifier is the element of the filter responsible for introducing the most distortion. While the output capacitance of the buffer/amplifier can be neglected, the input capacitance must be compensated during the design phase of the filter. In the band-pass filter, to facilitate the compensation of the amplifiers input capacitance, a new capacitance was added to the filters circuit, and the design equations were recalculated. During the simulation of the sixth-order low-pass filter it was determined that when using high quality factors, the filter is more sensitive to component variations and thus more susceptible to the components non-linearities.

Appendix A
Butterworth Filtering Transfer Function

A.1 Continuous-Time Low-Pass Butterworth Transfer Function

In order to obtain the values for the components in a filter, using the circuits transfer function, a prototype transfer function must be used. In this section it is described how to obtain the numeric values for the coefficients of the transfer function, using the Butterworth prototype transfer function. This transfer function can be obtained in several ways. It is possible to use a software tool, such as Matlab, to determine the prototype transfer function. In this case, it is necessary to indicate the order of the filter N and the desired cutoff frequency ω_p in [rad/s] (Table A.1), where n and d represent the values for the coefficients of the numerator and the denominator, respectively.

Alternatively, the filter can be designed using the mathematical equations first employed by S. Butterworth [18]. Using these equations, besides specifying the order of the filter and the cutoff frequency, it is also necessary to select the desired attenuation A_{max} at the cutoff frequency. The previous parameter is used to calculate the value of ε which is necessary to obtain the coefficients of the transfer function. The equation in Eq. A.1 shows how ε is calculated.

$$\varepsilon = \sqrt{10^{\frac{A_{max}}{10}} - 1} \qquad (A.1)$$

When the order of the filter is known, one of the equations in Eq. A.2 can be used to determine the denominator of the prototype transfer function [19], where B_n is the denominator of the normalized prototype transfer function, and $\hat{S}$ is the normalized

Table A.1 Obtaining the Butterworth transfer function coefficients using Matlab for a continuous-time low-pass filter

$[\text{n,d}] = \text{butter}(N, \omega_p, \text{'s'})$

© Springer International Publishing Switzerland 2015
H. A. de A. Serra, N. Paulino, *Design of Switched-Capacitor Filter Circuits using Low Gain Amplifiers,* SpringerBriefs in Electrical and Computer Engineering,
DOI 10.1007/978-3-319-11791-1

s plane.

$$
\begin{cases}
B_n(\hat{S}) = \prod_{k=1}^{\frac{n}{2}} \left[\hat{S}^2 - 2\hat{S} \cos\left(\pi \frac{2k+n-1}{2n}\right) + 1 \right] \text{ for even n} \\[2em]
B_n(\hat{S}) = (\hat{S} + 1) \prod_{k=1}^{\frac{n-1}{2}} \left[\hat{S}^2 - 2\hat{S} \cos\left(\pi \frac{2k+n-1}{2n}\right) + 1 \right] \text{ for odd n}
\end{cases}
\tag{A.2}
$$

A.1.1 First-Order Prototype Transfer Function

To implement a first-order filter, using Eq. A.2 with $n = 1$, the normalized prototype transfer function (Eq. A.3) is obtained.

$$
T(\hat{S}) = \frac{1}{\hat{S} + 1}
\tag{A.3}
$$

Depending on the type of filter (low-pass, band-pass, high-pass) the variable $\hat{S}$ is replaced with certain coefficients. To implement a low-pass filter, using Eq. A.3 with the low-pass coefficients, the first-order low-pass filter prototype transfer function (Eq. A.4) is obtained.

$$
T\left(\hat{S} = \frac{s}{\omega_p}\varepsilon^{\frac{1}{N}}\right) = \frac{\omega_p}{\omega_p + \varepsilon^{\frac{1}{N}}s}
\tag{A.4}
$$

A.1.2 Second-Order Prototype Transfer Function

To implement a second-order filter, using Eq. A.2 with $n = 2$, the normalized prototype transfer function (Eq. A.5) is obtained. This means that the quality factor for a second-order Butterworth filter is $Q_p = 1/1.4142 = 1/\sqrt{2} = \sqrt{2}/2$.

$$
T(\hat{S}) = \frac{1}{\hat{S}^2 + 1.4142\hat{S} + 1}
\tag{A.5}
$$

Using Eq. A.5 with the low-pass coefficients, the second-order low-pass filter prototype transfer function (Eq. A.6) is obtained.

$$
T\left(\hat{S} = \frac{s}{\omega_p}\varepsilon^{\frac{1}{N}}\right) = \frac{\omega_p^2}{\omega_p^2 + \frac{\omega_p \varepsilon^{\frac{1}{N}}}{Q_p}s + \varepsilon^{\frac{2}{N}}s^2}
\tag{A.6}
$$

Using either software tools or the Butterworth mathematical equations, the coefficients for the prototype transfer functions can be obtained. Since usually there are more components than equations, it is necessary to choose some component values in order to extract the remaining ones.

Table A.2 Obtaining the Butterworth transfer function coefficients using Matlab for a discrete-time low-pass filter

---

$[n,d] = \text{butter}(N, 2 f_p / F_s)$

Table A.3 Obtaining the Butterworth transfer function poles using Matlab for a discrete-time low-pass filter

---

$[z,p,k] = \text{butter}(N, 2 f_p / F_s)$

A.2 Discrete-Time Low-Pass Butterworth Transfer Function

The discrete-time Butterworth prototype transfer function can be obtained in a similar way. Using Matlab, it is necessary to indicate the order of the filter N and the relation between the cutoff frequency and the clock frequency $(2 f_p / F_s)$ (Table A.2), where n and d represent the values for the coefficients of the numerator and the denominator, respectively.

For higher order filters that are obtained from cascading lower order filter sections the code in Table A.2 is not convenient since it returns the coefficients of the higher order filter and not the coefficients for the cascaded sections. For this case Table A.3 can be used (where z, p, and k represent the coefficients of the zeros and of the poles, and the scalar gain, respectively) to obtain the poles of each section that can then be used to obtain the coefficients of each section.

Using the Butterworth mathematical equations, in order to convert the continuous-time prototype transfer function into a discrete-time prototype transfer function, a transform (bilinear, forward Euler, backward Euler) must be used.

A.2.1 First-Order Prototype Transfer Function

Using the forward Euler transform in Eq. A.4, the prototype forward Euler transfer function (Eq. A.7) is obtained.

$$T\left(s = \frac{1 - z^{-1}}{T_s z^{-1}}\right) = \frac{\omega_p T_s}{\omega_p T_s - \varepsilon^{\frac{1}{N}} + \varepsilon^{\frac{1}{N}} z} \tag{A.7}$$

When using the forward Euler transform, each continuous-time pole is converted into a discrete-time pole. Alternatively, the bilinear transform can be used, which transforms each continuous-time pole into a discrete-time pole and zero. Using this transform in Eq. A.4, the prototype bilinear transfer function (Eq. A.8) is obtained.

$$T\left(s = \frac{2}{T_s}\frac{z - 1}{z + 1}\right) = \frac{\omega_p T_s(1 + z)}{\omega_p T_s - 2\varepsilon^{\frac{1}{N}} + (\omega_p T_s + 2\varepsilon^{\frac{1}{N}})z} \tag{A.8}$$

Table A.4 Obtaining the Butterworth transfer function coefficients using Matlab for a continuous band-pass filter

---

[n,d] = butter(N/2,[ω_L ω_H],'bandpass','s')

A.2.2 Second-Order Prototype Transfer Function

Using the forward Euler transform in Eq. A.6, the prototype forward Euler transfer function (Eq. A.9) is obtained.

$$T\left(s = \frac{1 - z^{-1}}{T_s z^{-1}}\right) = \frac{T_s^2 \omega_p^2}{\varepsilon^{\frac{2}{N}} + T_s^2 \omega_p^2 - \frac{\omega_p}{Q_p} T_s \varepsilon^{\frac{1}{N}} + \left(\frac{\omega_p}{Q_p} T_s \varepsilon^{\frac{1}{N}} - 2\varepsilon^{\frac{2}{N}}\right) z + \varepsilon^{\frac{2}{N}} z^2}$$

(A.9)

Using the bilinear transform in Eq. A.6, the prototype bilinear transfer function (Eq. A.10) is obtained.

$$T\left((s = \frac{2}{T_s} \frac{z - 1}{z + 1}\right) = \frac{d(1 + z)^2}{a + bz + cz^2}$$

(A.10)

Where,

$$\begin{cases} a = 4\varepsilon^{\frac{2}{N}} + T_s^2 \omega_p^2 - 2\frac{\omega_p}{Q_p} T_s \varepsilon^{\frac{1}{N}} \\ b = 2T_s^2 \omega_p^2 - 8\varepsilon^{\frac{2}{N}} \\ c = T_s^2 \omega_p^2 + 2\frac{\omega_p}{Q_p} T_s \varepsilon^{\frac{1}{N}} + 4\varepsilon^{\frac{2}{N}} \\ d = T_s^2 \omega_p^2 \end{cases}$$

(A.11)

A.3 Continuous-Time Band-Pass Butterworth Transfer Function

The band-pass filter can be designed using the same methods described in Sects. A.1 and A.2. Table A.4 shows how to obtain the coefficients of the continuous time band-pass filter using Matlab, where N represents the order of the filter, ω_L and ω_H represent the lower and higher cut off frequency in [rad/s], and n and d represent the values for the coefficients of the numerator and the denominator.

Using the Butterworth mathematical equations and Eq. A.2 with n = 1, the prototype transfer function (Eq. A.12) is obtained. The prototype transfer function for a band-pass filter is designed with half the desired order since when replacing $\hat{S}$ with the band-pass coefficient, the order of the transfer function doubles.

$$T(\hat{S}) = \frac{1}{\hat{S} + 1}$$

(A.12)

Table A.5 Obtaining the Butterworth transfer function coefficients using Matlab for a discrete band-pass filter

---

$[n,d] = \text{butter}(N/2, [2 f_L/F_s \; 2 f_H/F_s], \text{'bandpass'})$

Using Eq. A.12 and replacing $\hat{S}$ with the band-pass coefficient, the second-order band-pass filter prototype transfer function (Eq. A.13) is obtained, where $Q_p = \omega_o/B$ and B is the pass band.

$$T\left(\hat{S} = \frac{Q_p\left(s^2 + \omega_o^2\right)\varepsilon^{\frac{1}{N}}}{\omega_o s}\right) = \frac{\omega_o s}{Q_p \omega_o^2 \varepsilon^{\frac{1}{N}} + \omega_o s + Q_p \varepsilon^{\frac{1}{N}} s^2} \tag{A.13}$$

A.4   Discrete-Time Band-Pass Butterworth Transfer Function

Using Matlab, it is necessary to indicate the order of the filter N and the relation between the cutoff frequencies and the clock frequency ($2 f_L/F_s$ and $2 f_H/F_s$) (Table A.5), where n and d represent the values for the coefficients of the numerator and the denominator, respectively.

The lower and higher cutoff frequencies can be calculated using Eq. A.14.

$$\begin{cases} f_H = f_o \left(\dfrac{1}{2Q} + \sqrt{\left(\dfrac{1}{2Q}\right)^2 + 1}\right) \\[4mm] f_L = f_o \left(-\dfrac{1}{2Q} + \sqrt{\left(\dfrac{1}{2Q}\right)^2 + 1}\right) \end{cases} \tag{A.14}$$

A.4.1   Second-Order Prototype Transfer Function

Using the forward Euler transform in Eq. A.13, the prototype forward Euler transfer function (Eq. A.15) is obtained.

$$T\left(s = \frac{1 - z^{-1}}{T_s z^{-1}}\right) = \frac{\dfrac{T_s \omega_o}{Q_p \varepsilon^{\frac{1}{N}}}(z - 1)}{1 + T_s^2 \omega_o^2 - \dfrac{T_s \omega_o}{Q_p \varepsilon^{\frac{1}{N}}} + \left(\dfrac{T_s \omega_o}{Q_p \varepsilon^{\frac{1}{N}}} - 2\right)z + z^2} \tag{A.15}$$

Using the bilinear transform in Eq. A.13, the prototype bilinear transfer function (Eq. A.16) is obtained.

$$T\left(s = \frac{2}{T_s}\frac{z - 1}{z + 1}\right) = \frac{d\left(z^2 - 1\right)}{a + bz + cz^2} \tag{A.16}$$

Where,

$$\begin{cases} a = -2T_s\omega_o + 4Q_p\varepsilon^{\frac{1}{N}} + Q_pT_s^2\omega_o^2\varepsilon^{\frac{1}{N}} \\ b = -8Q_p\varepsilon^{\frac{1}{N}} + 2Q_pT_s^2\omega_o^2\varepsilon^{\frac{1}{N}} \\ c = 2T_s\omega_o + 4Q_p\varepsilon^{\frac{1}{N}} + Q_pT_s^2\omega_o^2\varepsilon^{\frac{1}{N}} \\ d = 2T_s\omega_o \end{cases} \tag{A.17}$$

Appendix B
Impulse Response Simulation and Bode Diagram Plotting

In order to simulate SC filters impulse responses, the input signal must be a rectangular pulse that allows charge into the circuit, charging capacitors. Considering that charge enters the filter during clock phase ϕ_1, the rectangular pulse must occur shortly before phase ϕ_1 is 'high' and return to 'low' shortly after ϕ_1 returns to 'low' and before ϕ_2 begins.

In order to plot the Bode diagram from the impulse response, it is necessary to obtain a sample of the impulse response in every clock period until it stops oscillating. This information is then used in Matlab's *fft* function to convert the samples into complex numbers. These complex numbers can then be used to plot the Bode diagram. Figure B.1 graphically shows the procedure to obtain the transfer function from a switched capacitor filter.

An example of the code used to plot the Bode diagram from a SC filter is shown in Table B.1.

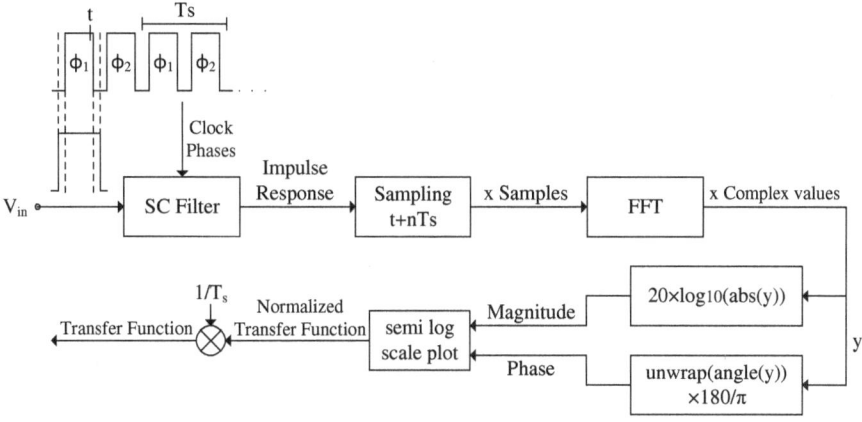

Fig. B.1 Procedure to obtain transfer function from a SC filter

© Springer International Publishing Switzerland 2015
H. A. de A. Serra, N. Paulino, *Design of Switched-Capacitor Filter Circuits using Low Gain Amplifiers*, SpringerBriefs in Electrical and Computer Engineering,
DOI 10.1007/978-3-319-11791-1

Table B.1 Example of how to plot the Bode Diagram

Freq = [0:n − 1]'/(n); *% n = Number of samples extracted from the impulse response*

data = **csvread**('filename.csv',1,0);

vout = data(:,2)/V; *% V = Amplitude of the input pulse*

time = data(:,1);

semilogx(Freq*Fs, 20***log10**(**abs**(**fft**(vout)))); *% Plots the magnitude*

semilogx(Freq*Fs, **unwrap**(**angle**(**fft**(vout)))*180/**pi**); *% Plots the phase*

References

1. T. Carusone, D. Johns, and K. Martin, *Analog Integrated Circuit Design*, ser. Wiley Desktop Editions. John Wiley & Sons, 2012.
2. J. T. Caves, S. D. Rosenbaum, M. A. Copeland, and C. F. Rahim, "Sampled analog filtering using switched-capacitors as resistors equivalents," in *IEEE J. Solid State Circuits*, vol. SC-12, no. 6, Dec. 1977, pp. 592–599.
3. R. Gregorian and G. Temes, *Analog MOS integrated circuitsfor signal processing*, ser. Wiley series on filters. Wiley, 1986.
4. A. P. Perez and F. Maloberti, "Performance enhanced op-amp for 65 nm CMOS technologies and below," in *Proc. IEEE Int. Symp. Circuits Systems (ISCAS'12)*, May 2012, pp. 201–204.
5. P. Allen and D. Holberg, *CMOS Analog Circuit Design*, ser. The Oxford Series in Electrical and Computer Engineering. Oxford University Press, USA, 2002.
6. B. J. Hosticka, R. W. Brodersen, and P. R. Gray, "MOS sampleddata recursive filters using switched capacitor integrators," in *IEEE J. Solid State Circuits*, vol. SC-12, no. 6, Dec. 1977, pp. 600–608.
7. K. Martin, "Improved circuits for the realization ofswitched -capacitor filters," in *IEEE Trans. Circuits and Systems*, vol. SC-27, no. 4, Apr. 1980, pp. 237–244.
8. R. P. Sallen and E. L. Key, "A practical method of designing RC active filters," in *IRE Trans. Circuit Theory*, vol. CT-2, Mar. 1955, pp. 74–85.
9. H. Serra, N. Paulino, and J. Goes, "A switched-capacitor biquadusing a simple quasi-unity gain amplifier," in *Proc. IEEE Int. Symp. Circuits Systems (ISCAS'13)*, May 2013, pp. 1841–1844.
10. M. Dessouky and A. Kaiser, "Input switch configuration suitablefor rail-to-rail operation of switched opamp circuits," in *Electronics Letters*, vol. 35, no. 1, Jan. 1999, pp. 8–10.
11. F. Michel and M. Steyaert, "A 250 mV 7.5 μW 61 dB SNDR SC $\Delta\Sigma$ modulator using near-threshold-voltage-biased inverter amplifiers in 130 nm CMOS," in *IEEE J. Solid-State Circuits*, vol. 47, no. 3, Mar. 2012, pp. 709–721.
12. B. Razavi, *Design ofAnalog CMOS Integrated Circuits*. New York, USA: McGraw-Hill, Inc., 2001.
13. A. D. Grasso, G. Palumbo, and S. Pennisi, "Three-stage CMOS OTA for large capacitive loads with efficient frequency compensation scheme," in *IEEE Trans. Circuits and Systems II: Express Briefs*, vol. 53, no. 10, Oct. 2006, pp. 1044–1048.
14. L. Acosta, R. G. Carvajal, M. Jimenez, J. Ramirez-Angulo, and A. Loper-Martin, "A CMOS transconductor with 90 dB SFDR and low sensitivity to mismatch," in *IEEE Int. Symp. on Circuits and Systems (ISCAS 2006)*, May. 2006, pp. 69–72.
15. F. Krummenacher and N. Joehl, "A 4-MHz CMOS continuous-timefilter with on-chip automatic tuning," in *IEEE J. Solid-State Circuits*, vol. 23, no. 3, Jun. 1988, pp. 750–758.
16. K.-C. Kuo and A. Leuciuc, "A linear MOS transconductor usingsource degeneration and adaptive biasing," in *IEEE Trans. Circuits and Systems II: Analog and Digital Signal Processing*, vol. 48, no. 10, Oct. 2001, pp. 937–943.

© Springer International Publishing Switzerland 2015

H. A. de A. Serra, N. Paulino, *Design of Switched-Capacitor Filter Circuits using Low Gain Amplifiers,* SpringerBriefs in Electrical and Computer Engineering, DOI 10.1007/978-3-319-11791-1

17. H. Serra, N. Paulino, and J. Goes, "A switched-capacitorband -pass biquad filter using a simple quasi-unity gain amplifier," in *Technological Innovation for the Internet of Things (DoCEIS'13)*, 2013, vol. 394, pp. 582–589.

18. A. S. Sedra and K. C. Smith, *Microelectronic Circuits*, ser. The Oxford Series in Electrical and Computer Engineering. Oxford University Press, USA, 2004.

19. R. M. Golden and J. F. Kaiser, "Root and delay parameters fornormalized bessel and butterworth low-pass transfer functions," in *IEEE Trans. Audio and Electroacoustics*, vol. AU-19, no. 1, Mar. 1971, pp. 64–71.